ELECTRICAL COUNTING

ELECTRICAL COUNTING

With special reference to counting

ALPHA AND BETA PARTICLES

by

W. B. LEWIS, M.A., Ph.D.

Fellow of Gonville and Caius College, Cambridge
University Lecturer in the Cavendish Laboratory

CAMBRIDGE

AT THE UNIVERSITY PRESS

1942

CAMBRIDGE
UNIVERSITY PRESS

University Printing House, Cambridge CB2 8BS, United Kingdom

Cambridge University Press is part of the University of Cambridge.

It furthers the University's mission by disseminating knowledge in the pursuit of education, learning and research at the highest international levels of excellence.

www.cambridge.org
Information on this title: www.cambridge.org/9781316611760

© Cambridge University Press 1942

First published 1942
First paperback edition 2016

A catalogue record for this publication is available from the British Library

ISBN 978-1-316-61176-0 Paperback

CONTENTS

PREFACE

The technique of electrical counting has grown as an essential aid in research in nuclear physics and this book owes much to those who have pursued this science at the Cavendish Laboratory.

The central chapters of this book, beginning with Chapter III, will I hope be of interest to many who have occasion to use valve circuits, but who may not be concerned with nuclear physics. The application of the technique to nuclear physics is unfortunately somewhat in abeyance owing to the war at the time of publication. I must also plead the war as my excuse for not having included references to work published since the beginning of the war.

I wish to express my thanks to all those of the Cavendish Laboratory who have assisted with the preparation of this book and especially to Dr N. Feather and Mr J. A. Ratcliffe who have read and made valuable contributions to the text, and to Mr G. R. Crowe who has drawn a number of the figures. The fast counting meter illustrated on page 80 owes much in design and construction to Mr J. Fuller of the Cavendish Laboratory workshops.

W. B. LEWIS

1942

Chapter I

IONIZATION BY SINGLE PARTICLES

Individual atomic and subatomic particles may be detected if they are electrically charged and possess sufficient kinetic energy. When such particles pass through matter the electric charge produces mechanical forces sufficient to disrupt the electronic configurations of the atoms through which they pass. A trail of charged atoms or ions is therefore left in the wake of the swiftly moving particle.

One method of observing this trail of ions is by its electrical effect, and it is with this general method that this book is exclusively concerned. Another important method of observation is provided by the Wilson expansion chamber in which visible water drops may be condensed on the ions of such a trail in a gas. These two methods are the most valuable means of observation known in nuclear physics. They are closely interdependent; expansion chamber observations facilitate and make more certain the deductions from the electrical effects. The expansion chamber is supreme for observations in three dimensions. The electrical method, on the other hand, is unsurpassed for the rapid accumulation of data and for exact energy measurements. Two further methods are also of importance, namely, observation of the photographic effect of the trail of ions in a sensitive photographic emulsion, and of the luminous effect of the trail in certain crystals such as zinc sulphide.

Electrical methods of counting the particles all depend on the ionization of the trail. It is therefore important to know the ionization per unit length, usually called the specific ionization, produced by different particles in different media. This specific ionization depends to a certain extent on the energy of the particle. Except for high-energy particles of energy greater than 10^8 e.V., such as are encountered in cosmic rays, there exists experimentally a sharp division between the light particles which are positive and negative electrons (β-rays), and the heavy particles such as protons, α-particles and other atomic nuclei. The latter produce such dense ionization that it is possible to measure, at least roughly, the ionization in less than a millimetre of a single track in air at atmospheric pressure. The total range in air of a typical α-particle of high energy (~ 7 M.e.V.) from a radio-active substance would be a few centimetres, and the trail would contain more than a hundred thousand ions. The specific ionization on an electron track, on the other hand, is so slight (4–50 ions/mm. air) that only by some method such as amplifying

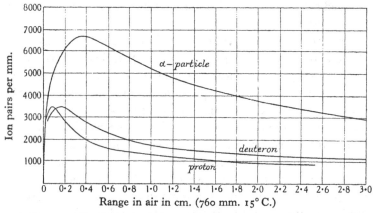

Fig. 1·1. Specific ionization. Range curves for different particles.

the ionization by collisions can the ionization on a single electron track be measured. Though the ionization is not easily measured it may readily be detected by a device in which it releases a momentary electrical discharge. Such discharge counters, as they are termed, may be so sensitive that a discharge is produced by a single pair of ions.

Fig. 1·1 shows specific ionization-range curves for the commoner particles in air. Except for ranges of a few millimetres these curves are approximately of the same shape but with the scale of ranges proportional to M/Z^2, where M is the mass and Z the charge number of the particle, and the ionization scale proportional to Z^2. At short ranges the phenomenon by which the effective charge of an α-particle is reduced by capture of an electron disturbs this relation. The curves shown are based on experimental observations. For long ranges the ionization decreases approximately as $\sqrt[3]{1/R}$, where R is the range, and for still longer ranges as $\sqrt[4]{1/R}$.

Knowledge of the specific ionization, besides being important for a proper appreciation of electrical recording processes, has also led to means of determining the energy of a particle with a high degree of precision. The energy of such a subatomic projectile is always ultimately determined by the curvature of its path in a magnetic field together with determinations of its mass and charge. It is, however, found that the ionization accounts for practically all of the energy lost by the projectile. Now if the specific ionization and also the energy loss in forming a pair of ions were constant for a particle of any energy, it would follow that the particles would travel a total distance or range dependent only on the initial energy. This is found to be approximately true, but observation shows that particles initially of the same energy have ranges which are not exactly equal. This is explained as due to the statistical nature of the ionization process. This straggling of the ranges of particles of a given energy has been studied in detail for α-particles. Consequently it is now possible to determine the initial energy of a group of α-particles with a very high degree of precision by measuring the ranges.

In common with all accurate measurements such α-particle range measurements require careful attention to many details. These will be discussed further in the last chapter, but here it may be stated that such measurements require the measurement of the ionization over a certain small length of the track of individual particles. Chapter XII is included in the present account just because a discussion of the means for measuring this ionization will be one of our main concerns.

The tracks of the heavy particles through air are for the most part straight, only occasional large-angle deflexions occurring. These are most common in the last few millimetres of the range. The last few centimetres of the track of an electron, on the other hand, are so curved that measurements of individual ranges can only be made with any certainty in an expansion chamber. Measurements of electron ranges by electrical recording result in limiting ranges from the source which are considerably shorter than ranges measured along the track. Nevertheless, such measurements are of value if made in some dense absorber such as a solid so that the lateral deviation of the rays is minimized. It is, moreover, possible to determine energies from the ranges with some precision. For the high-energy electrons for which this method is most suitable the range varies almost linearly with the energy.

Chapter II

COUNTING IONIZATION CHAMBERS

The function of a counting ionization chamber is to produce a change of potential on the input grid of the first valve of the amplifier from the ionized track in the chamber. Usually the chamber is operating with such small total charges that the maximum changes of potential are not many times greater than the minimum it is possible to detect. It is therefore desirable to collect as much of the charge as possible on an electrode having a minimum of capacity. Little advantage will, however, be gained by making this capacity much less than the inherent valve input capacity which is generally of the order of a few centimetres.

The size and shape of the chamber is usually determined by its particular function; before, however, discussing special designs a few general considerations may be noted. One electronic charge e on a capacity of C cm. produces a potential change $300e/C$ V. $= 0.144/C$ μV. Under ordinary circumstances it is possible to detect on a capacity of 10 cm. a charge of $500e$, which produces a change of potential of 7.2 μV. Such a small change of potential is only detectable if it takes place suddenly; this is explained at the end of this chapter and in the next on the limitations of amplifiers. It is found that the time of collection of the ions should be as short as possible, but an alteration of the amplifier or recording system may be required to take advantage of a very short collecting time, and with a given amplifier it is possible that no advantage will be gained by shortening the collecting time beyond a certain point.

The insulation of the collecting electrode must be of the highest order. Leakage across good insulators takes place mainly over the surface. Currents flowing in thin surface films are liable to be unsteady; this gives rise to fluctuations of charge which may affect directly or by electrostatic induction the potential of the collecting electrode. It is not the magnitude of the insulation resistance which is important, but the nature of the path taken by the leakage currents.

The collecting electrode should be supported by insulators only of the highest grade, such as quartz, amber, sulphur or sealing wax. The insulators need not be long, for a variable surface charge cannot be tolerated. If the surface is bad the insulator is useless. The fixed ends of the insulators should be attached to metal which is as near as possible to the potential of the collecting electrode. The insulators should preferably be placed so that no charge is driven on to them from the ionization in the chamber; satisfactory chambers have, however, been made in which this precaution has not been observed.

The insulation of the high-potential electrode need not be of such a high order; rubber and ebonite are satisfactory. The insulation should be placed so that a variable surface charge on it does not induce charges on the collecting electrode. If the high-potential electrode is not completely screened it is liable to pick up stray electromagnetic disturbances. To minimize the effect of this the electrode should have a large capacity to earth. The condenser providing this capacity should be placed very close to the high-potential electrode and the earth connexion should be as short as possible because any length of wire has an appreciable reactance for some sufficiently high frequency. Such high-frequency disturbances are not in themselves harmful, but when rectified by the valve input circuit they give rise to disturbing low-frequency impulses.

When it is desired to have a very high potential difference across the chamber, trouble may be experienced from small discharges of the corona type occurring at the junctions between the solid insulator and the metal. This phenomenon is caused by the difference of dielectric constant between the insulator and the gas in the chamber; it is therefore desirable to use an insulator with as low a dielectric constant as possible. Sulphur and resin waxes are good in this respect, with dielectric constants of 3–4. The conductor should also be in intimate contact with the dielectric; sulphur and the waxes are again very suitable as they may be melted on to the metal. For potentials over 2000 V. it may be necessary or more satisfactory to subdivide the insulation, using a number of metallic guard rings maintained at intermediate potentials by resistances. The resistances must be very constant unless bridged by condensers. The same considerations apply to the insulation of the condenser connected across the chamber to prevent electromagnetic pick-up.

It is sometimes necessary to mount the first valve at some distance from the ionization chamber. This may happen when the ionization chamber must operate in a strong magnetic field. The most satisfactory means of connecting the collecting electrode to the first valve is probably by a thin wire along the axis of a wide metal tube. This keeps the capacity low but provides a large volume from which ions may be collected. To avoid this collection of ions the wire should be at the same potential as the surrounding tube. Where a difference of potential of only a volt or so occurs it has been found satisfactory to run the wire through a piece of narrow quill glass tubing along the axis of the wide metal tube. The insulation at the ends of the tube may be improved with sealing wax to reduce the leakage of charge from the glass to the collecting electrode. It is presumed that the outer surface of the glass tube quickly takes up such a potential that there is no further collection of charge. The quill tube by preventing movement of the wire also minimizes microphonic effects.

Ionization chambers are very liable to be microphonic; this is one of the most important points to be considered in the design.

For the accurate measurement of ranges the particles should enter the chamber through a uniform or very thin metallic foil. Ions must be collected from the space immediately adjacent to the foil, so the foil is necessarily in a strong electric field. Slight movements of the foil are therefore liable to induce charges on the collecting electrode and the chamber is consequently microphonic.

The design shown in Fig. 2·1 probably represents the simplest chamber satisfactory for accurate range measurements. The collection of ions from all tracks penetrating the same distance is exactly similar, so the amount of the charge collected is a measure, although ambiguous, of the depth of penetration. The ambiguity arises because of the maximum in the ionization-range curve for a single particle. An α-particle which passes right through the chamber and still has a residual range of about 2 mm. after reaching the back wall of the chamber will give most ionization. A particle which does not penetrate so far, as also a swifter particle which has still farther to go, will give less ionization.

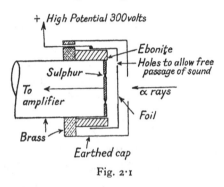

Fig. 2·1

The ions are separated by the electrostatic field in a direction approximately along the track. The ions will not be all collected at the same time. If the ionization along the track is approximately uniform, the rate of collection of ions will be also approximately uniform. The potential of the collecting electrode will not, however, rise uniformly, because before the ions are collected the potential will be altered by the induced charge. In an exact calculation of this induced charge it would be necessary to consider the infinite series of electrical images, the inhomogeneity of the ionization, the different ionic mobilities, the negligible change of potential of the guard ring, and the total capacity of the collecting electrode. Here, however, a very simple consideration must suffice. If Fig. 2·2 represents positive ions moving to the right and negative ions to the left under the action of the field between the high-potential electrode and the collecting electrode, and if the ions close to either electrode are assumed effectively to have been collected, then it will be seen that the initial rate of effective collection of ions is just double the rate of arrival of ions at the electrode. When only a few ions are left in the chamber these will all be close to one or other of the two electrodes and the effective rate of collection will be zero.

It has been mentioned that the time of collection of the ions should generally be as short as possible. This time may be obtained from a knowledge of the electric field and the mobilities of the ions. The mobility of positive ions in air at atmospheric pressure is about 1·35 cm./sec./V./cm.; the mobility of negative ions

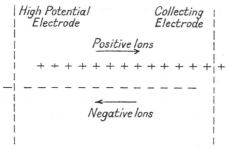

Fig. 2·2

depends on the moisture and may be taken as 1·5 in moist air and 1·8 in dry air. Mobilities may be assumed to be inversely proportional to the pressure, but at pressures less than about one-tenth of an atmosphere the mobility of negative ions increases much more rapidly as the pressure is reduced. The mobility of ions in hydrogen at atmospheric pressure is approximately 6 cm./sec./V./cm.

Hydrogen may with advantage be used instead of air as the gas in an ionization chamber, for the high mobilities of ions and the relatively low stopping power make it possible to use deeper chambers and thus diminish the effects of small movements of the foils, so the chamber is less microphonic. If the chamber cannot be sealed gas-tight, it has been found possible to work with a steady stream of hydrogen flowing through.

Special chambers.

A chamber which is not affected by ordinary noises may be made by making the high-potential cap of thick (> 1 mm.) brass and cutting a grid in this for the particles to pass through. The mesh of the grid should not be greater than the thickness of the brass; the holes may be square to reduce the number of particles stopped. The outer surface may then be covered with thin gold or aluminium leaf (stopping power 0·4–1 mm.).

This type of chamber is unsatisfactory if the particles are already passing through a grid, as usually happens when the source must be in a vacuum, for then it is difficult to move the chamber without altering the relative shadow ratio of the two grids. The number of particles counted at a given range is thus variable.

Differential chamber.

For the highest accuracy in the measurement of ranges a differential chamber (Fig. 2·3) is used. The collecting electrode in this is a thin foil through which the particles pass. The electric

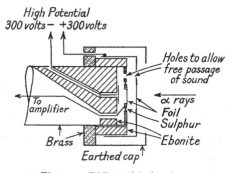

Fig. 2·3. Differential chamber.

field on either side of this foil is in the same direction in space, so that positive ions are collected on one side and negative ions on the other. By suitably proportioning the depths of the two chambers it may be arranged that a particle passing right through the two chambers gives only a very small nett charge to the collecting electrode. The size of the impulse as a function of the residual range of an α-particle entering the front chamber is shown for a typical differential chamber in Fig. 2·4. It should be realized that the form of this curve depends on the specific ionization of the particle at points near the end of its range; this

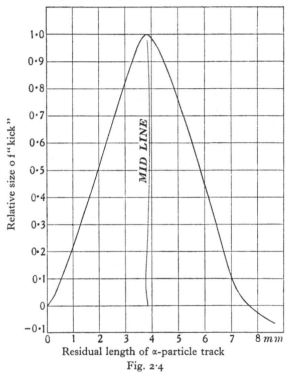

Fig. 2·4

Relative size of "kick" (y-axis)

MID LINE

Residual length of α-particle track

is, unfortunately, not known with accuracy, as explained in Chapter I.

It will be realized that a differential chamber is very microphonic. If possible the source and chamber should be sealed up in an airtight box to exclude sound, and the whole should be mounted on rubber or some sound-absorbing material to avoid the transmission of vibrations by contact. Where this is not possible it has been found best to go to the other extreme and make the chamber as much of an openwork structure as possible, rings of large holes being drilled through the disks supporting the foils, and through the sides of the tubes to which the disks are fitted. The object is to prevent as far as possible differences of pressure from existing between the two sides of the foil. It should, however, be arranged that the forms of the collecting and high-potential electrodes are not such that they would ring even at a very high pitch when struck.

Discriminating chamber.

Alfvén [2] has shown that it is possible to obtain a rough trace of the distribution of ionization along the track in an ionization chamber and thus to discriminate between α-particles, protons and possibly deuterons, triplons and He^3 particles. For this it is necessary to avoid the effect of the charge induced on the collecting electrode before the actual collection of the ions. This is achieved (Fig. 2·5) by placing the collecting electrode behind a grid which

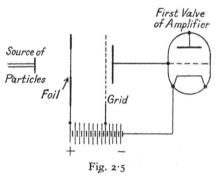

Fig. 2·5

replaces the collecting electrode in an ordinary single chamber such as has been described. The potentials are arranged so that in the ionization chamber proper the drift velocity is smaller than in the chamber behind the grid. A certain proportion of the ions arriving at the grid pass through; these are quickly drawn to the collecting electrode, so the rate of collection of ions is proportional to the rate of arrival at the grid. This in turn is proportional to the ionization density of that portion of the track which is at that instant arriving at the grid.

The writer has found that by a suitable design of coupling circuit in the amplifier it is possible to obtain from such a chamber an impulse of the form shown in Fig. 2·6. On an oscillograph record with the paper running at a high speed this would have the form shown at (*a*), and with the paper running slowly this has the form (*b*). The height of the thick portion is a measure of the total ionization, and the extra height of the thin portion is a measure of the ionization density at the front of the chamber. This allows a discrimination between α-particles and protons which is useful when protons must be counted in the presence of large numbers of α-particles. For the same total ionization the proton will give less ionization at the front of the chamber.

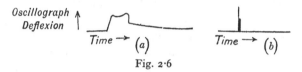

Fig. 2·6

A similar discrimination between particles may be made by using a deep chamber to avoid the ambiguity of the interpretation of the size of an impulse as a measure of the penetration distance. To avoid the slow collection of ions along the depth of a deep chamber it is better to let the particles travel parallel to the electrodes which may then be closer together. By measuring the variation of the average size of impulse as the distance from the source is altered, the ionization density along the track may be determined and hence the particle identified.

Ionization chambers for recording α-particles and protons in the presence of intense β- and γ-radiation.

The simple chamber and the differential chamber already described were developed primarily for the observation of α-particles in the presence of a background of ionization much more intense than the ionization due to the α-particles. It proved possible to work with an ionization which on the average was 100,000 times the ionization due to the α-particles. Such discrimination was possible because the ionization due to a single α-particle was formed practically instantaneously and the charge collected in about 2×10^{-4} sec. The background ionization might be the equivalent of that from 10^5 α-particles per sec., so that in 2×10^{-4} sec. it would be equivalent to 20 α-particles or say a total of 400,000 ions from a chamber 3·5 mm. deep. This rate is, however, continuous except for the statistical fluctuations which would amount to $\sqrt{(n \Delta t)}$ ions in a short time Δt, where n is the average number of ions/sec. The probable fluctuation in 2×10^{-4} sec. would thus be only $\sqrt{(400,000)} \simeq 630$ ions. The arrival of an α-particle produces a charge of 20,000 ions, which could readily be distinguished from the background fluctuations.

As explained in the next chapter this background is added to by that arising in the amplifier. The relative importance of this may be decreased by increasing the ionization collected by allowing proportional amplification of the ionization by ionic collisions, which takes place in strong electric fields. Such strong electric fields also decrease the time of collection of the ions, which enables the background to be further reduced. This practice of obtaining proportional amplification by collision has been successfully used in experiments where protons were to be observed at a slow rate with a large background of γ-ray ionization, as in experiments on artificial disintegration produced by the α-particles from thorium C and C′, and radium C′.

Proportional amplifying chambers.

Geiger and Klemperer[25] studied in detail the behaviour of counters in which the negative electrode is a small polished sphere. Many workers have since used this type of counter with which it is possible to obtain a proportional amplification of the charge up to about 10^4 as a limit, but more usually an amplification of about 10^3 or less is employed, as such amplification is more stable and consistent.

Zipprich[82] has developed a chamber with parallel plane electrodes, one of which is a grid through which the ions are admitted to the main accelerating space. The electric field in this space penetrates to some extent through the meshes of the grid so that ions are drawn to the grid and some pass through. A proportional amplification of about 1000 was used.

When the voltage on a proportional counter is raised the amplification tends to become unstable, and for an amplification exceeding 10^5 it is found that discharges are set up independent of the initial ionization. This process has been developed and is widely used for recording β- and γ-rays. The Geiger-Müller counter discussed in Chapter IX operates on this principle.

Chapter III

LIMITATIONS OF AMPLIFIERS

When the amplification or gain of an amplifier is made very great a background is observed. On listening with earphones to the output of a high-gain audio-frequency amplifier this background is heard as a rushing noise. Three main sources of this background noise have been recognized. First, noise dependent on the input circuit but independent of the current in it; this is identified as an effect of thermal agitation. Secondly, noise which increases with the anode current of the first amplifying valve; one recognized cause of such noise is known as the shot effect. Thirdly, when a current flows through certain types of conductor it is presumed that the lines of current flow do not remain constant in detail, and this gives rise to noise generally rather more irregular than noise from the first two sources but sometimes difficult to distinguish. For example, if the current is accompanied by electrolysis which results in the formation of gas at the electrodes, the lines of current flow will constantly be interrupted by the bubbles of gas. A somewhat similar effect is produced when current flows through thin films of metal or carbon. This depends on the physical nature of the film, for carbon film resistances which do not show this effect are made and marketed by Siemens-Schuckert[11,29], but the writer has found no single specimen of any other of the many film resistances, commonly advertised as silent, which is free from this effect. Carbon composition resistances are similarly defective. Current-carrying contacts with insufficient contact pressure similarly give rise to noise. The cathodes of certain valves show a similar effect; one particular phenomenon of this type has been named the "flicker" effect[36], the emission from any particular point of a cathode appearing to flicker though the emission of the whole may appear constant. This effect is most apparent at frequencies below about 1000 c./sec.

The third source of noise may be eliminated by using only well-made wire-wound resistances (or the Siemens-Schuckert resistances) in the early stages of the amplifier, and by selection of the first valve. All contacts must have sufficient contact pressure and sufficient contact area for the current they have to carry.

The magnitude of the thermal agitation effect may be derived directly from the principle of the equipartition of energy. The idealized elementary circuit (Fig. 3·1) has only one natural or characteristic frequency; it thus has only two degrees of freedom and we may write $\frac{1}{2}Li^2 = \frac{1}{2}CV^2 = \frac{1}{2}kT$, where i^2 is the mean square current flowing and V^2 the mean square voltage on the condenser. k is Boltzmann's constant, T is the absolute temperature. In order to apply this we require to know the distribution of this energy over the frequency spectrum. This problem is exactly analogous to the de-

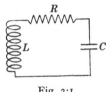

Fig. 3·1

termination of the distribution of radiant energy in the black-body spectrum. The calculation on these lines was first made by Nyquist (53). Although this is very illuminating it is a little too long to be included here, and we may just note the conclusions. First it is to be noted that the thermal energy is communicated to the idealized circuit through its resistance R. If then we inquire the rate at which two equal resistances interchange energy as a function of frequency, it is possible to determine the thermal agitation e.m.f. existing between the ends of a pure resistance as a function of frequency. This resistance with its thermal e.m.f. may then be considered inserted in the elementary oscillatory circuit and the resultant voltage across the condenser may be calculated.

If E_ν^2 is the mean square e.m.f. between frequencies ν and $\nu + d\nu$ existing between the ends of a resistance R at temperature T, then it is found that $E_\nu^2 = 4RkT\,d\nu$. Calculating V_ν^2 from this and integrating over all frequencies leads to the expected result $V^2 = kT/C$, which we note is independent of R. The frequency distribution does, however, depend on R.

It may be noted that $\int_{\nu=0}^{\nu=\infty} E_\nu^2$ for a pure resistance is infinite. This is the old impasse of the classical theory of the distribution of energy in the black-body spectrum and is resolved, as in that case, by the adoption of quantum principles of the partition of energy at high frequencies. On the other hand, when the resistance e.m.f. is applied to our idealized elementary circuit no resultant voltage is found across the condenser at high frequencies. This paradox is again resolved by noting that no real circuit has only one characteristic mode of oscillation; it has $3n$, where n is the number of atoms in the circuit (cf. Debye's [17] theory of the specific heats of solids).

In the particular application to amplifiers for ionization pulses we are concerned with thermal agitation in the input circuit at low frequencies up to 1 or 2×10^4 c./sec. At these frequencies the circuit may be considered simply as a condenser C shunted by a resistance R for which we have

$$V^2 = \frac{2RkT}{\pi} \int_{\omega=0}^{\omega=\infty} \frac{d\omega}{\omega^2 C^2 R^2 + 1} = \frac{2kT}{\pi C} \left[\tan^{-1} \omega C R \right]_0^\infty$$

If $C = 10\,\mu\mu$F. and $R = 10^{10}$ ohms, $CR = 0\cdot1$ sec. Frequencies higher than 10 c./sec. $(\omega C R > 2\pi)$ contribute only 10 % to V^2.

It is rarely necessary to amplify frequencies as low as 10 c./sec., and ordinarily thermal agitation in the input circuit may be neglected.

Thermal agitation does, however, provide a convenient means of calibrating amplifiers, for if R is reduced to 10^5 ohms, then $CR = 10^{-6}$ sec., so that the thermal agitation noise covers the frequency range amplified, and the sensitivity of the amplifier may be gauged by measuring with a thermo-couple or other square law device the rushing noise in the output circuit when a 100,000-ohm resistance is connected between the first grid and cathode. If the frequency range of the amplifier is known, this may be made the basis of an absolute calibration of the amplifier. Even without this knowledge the method is found useful for comparing similar amplifiers and for keeping a check on amplification and background level.

As has already been stated the background due to thermal agitation in the input circuit of a pulse amplifier should be negligible. The main contribution to the background is provided by the first valve itself. Many partial explanations of this background have been discovered, but it has not been found possible to predict its magnitude from fundamental principles. Some of the recognized contributory phenomena may be mentioned. First the valve possesses an internal resistance, which means that the energy dissipation in the volume distribution of charge within the valve is controlled by the applied potentials. The statistical thermal variations in this space charge conversely affect the electrode potentials. This space charge is, however, very far from thermal equilibrium. It is thus not possible in any simple way to assign a magnitude to the thermal agitation voltage, but on general grounds it might be supposed that the voltage would be greater the higher the temperature of the cathode. Evidence in support of this has been obtained, notably by Pearson (54). At low frequencies, however, the flicker effect predominates.

Particularly at high frequencies a contribution from another fundamental cause may predominate. This is known as the shot effect and depends on the finite size of the electronic charge. It must be emphasized that this effect is never fully operative in an amplifying valve. Its magnitude may, however, be quite simply derived for a valve worked under saturation conditions, under which the arrival of electrons at the anode is independent of small variations of anode potential, and in which the time of transit of an electron from cathode to anode is negligible compared with the periodic times of the frequencies concerned. Let C be the anode-cathode capacity and R the external anode circuit resistance. Let Q be the instantaneous charge on the anode and $\bar{i}$ the mean anode current. The energy E stored in C is then $\frac{1}{2}Q^2/C$, but only discrete changes of Q of amount e are possible, where e is the electronic charge, so we have $E + \Delta E = (Q+e)^2/2C$ or $\Delta E = eQ/C + e^2/2C$. If n electrons reach the anode in unit time the total change of energy in unit time is $n\Delta E = neQ/C + ne^2/2C$. This energy is dissipated in R, but the energy dissipated in R in unit time is $\bar{i}^2 R + i_f^2 R$, where i_f = fluctuation current. But $\bar{i} = ne$; therefore $neQ/C = \bar{i}Q/C = \bar{i}^2 R$. So we have $i_f^2 R = ne^2/2C$, or $i_f^2 = \bar{i}e/2CR$, which is the well-known shot effect equation. This derivation is due to E. B. Moullin [49].

The mean square fluctuating voltage across R is therefore proportional to e, the electronic charge, and $\bar{i}$, the mean anode current. As an example of the magnitude, the root mean square voltage = 200 μV. for $\bar{i} = 1$ mA., $C = 20\,\mu\mu$F., $R = 10^5$ ohms. The energy is then spread over frequencies of the order $1/CR = 500,000$ radians/sec.

When as in an amplifying valve the anode current depends on the anode potential, these fluctuations tend to be smoothed out, for if in a certain interval too few electrons are received the anode potential rises and thus increases the mean current. But this is not the only effect operating to reduce the shot effect. The anode current is limited by the presence of an electronic space charge round the cathode; the transit of an electron from cathode to anode cannot therefore be considered to occupy a negligible time. An electron in the space between cathode and anode induces charges on the electrodes which vary with the position of the electron. The change of energy of the anode-cathode capacity cannot therefore be considered as taking place abruptly by amount e.

In practice the selection of the first valve and its operating conditions are determined by experiment. A few general conclusions may be noted: an uncoated tungsten filament valve is liable to give occasional clicks and the general noise level is liable to be high; a thoriated filament may be satisfactory, but valves with such filaments are now rare and are liable to be microphonic; the grid insulation of modern valves with oxide-coated cathodes is often unsatisfactory, particularly if the electrodes are positioned by mica spacers the surface path over which between electrodes is short; defective grid insulation may give rise to noise even though the insulation resistance may be high;

the noise level of most valves increases relative to impulses on the grid when the anode potential is increased above about 15 V., this being presumably due to ionization; there are some exceptions to this rule.

No valve specially designed for the purpose appears to be on the market yet. Perhaps the nearest to a special design is the Osram A 537, designed as the input valve for working from a photoelectric cell for reproducing sound from film. This is a simple indirectly heated triode with an amplification factor of 15·5. Special attention has been paid to the grid insulation, not forgetting the avoidance of microphony. This is achieved by bringing the grid lead out separately at the top of the valve and by using steatite in place of mica spacers to hold the electrodes.

The input valve of an amplifier for ionization pulses receives a certain charge on the grid, and it is desired that the resultant change of grid potential should be large. For this to be so the effective grid input capacity must be as low as possible. It is explained in the next chapter that this is more easily reduced by using a screen-grid construction, leading to a tetrode or pentode instead of a triode.

Many screen-grid valves are now made with the grid lead brought out separately at the top of the bulb, and in some of these consideration has been given to the grid insulation, slots being cut in the mica spacers to increase the surface path over the mica. Selected specimens of these are as satisfactory or better than the A 537, and as these valves are used in ordinary radio sets their price is much lower. The best valve that the writer has yet encountered is a Western Electric type 310 A tetrode. This, however, has a filament rating 10 V. 0·3 amp. which is generally inconvenient. The Mullard Type SP 4 B pentode has been used successfully. All valves with oxide-coated cathodes are more silent or not less silent if the cathodes are run at a lower temperature than that normally recommended by the makers, and the filament voltage may generally with advantage be from 10 to 20 % lower than the maker's rating.

Some valves (particularly triodes) of very high amplification factor, such as Osram H 42 or H 30, cannot be operated with the grid isolated, since the grid charges up to such a negative potential that the anode current at low-anode voltages is reduced to zero. A leak resistance of about 10^9 ohms may be connected from the grid to cathode to avoid this without causing trouble except when very low frequencies are to be amplified.

The Osram D.E.V. has been used a great deal in this work. This is a small valve with a thoriated tungsten filament in which the grid is of low capacity and supported from a glass stem well away from the anode and filament supports. The disadvantage of this valve is that it is microphonic. Its noise level is, however, low, though not lower than that of some of the new valves such as the Osram A 537. Microphonic trouble with the D.E.V. is lessened up to a point by increasing the filament temperature. Also it will generally be found that the filament is not quite concentric with the grid and anode. The valve is least microphonic when mounted with the filament horizontal and in such a position that the sag of the filament brings it nearest to the centre of the electrode system.

It is necessary to consider how a given change of potential on the grid of the first valve may be made to stand out above the background fluctuations. The potential V of the grid of the first valve is a continuous function of time, and its variation is brought about by the statistical fluctuations of a large number of events. The probability (if small) of a certain change ΔV taking place in a time ΔT will be proportional to ΔT. If, therefore, an impulse which is a given small change of V is to be distinguished from the background it must take place in a short time. How such an impulse may be separated from the background is discussed in the chapter on the design of amplifiers, and further in the section on discriminators.

Chapter IV

DESIGN OF AMPLIFIERS

Simple theory.

The simplest element of a resistance-capacity coupled amplifier is shown in Fig. 4·1. In this C_c is the coupling condenser, R the

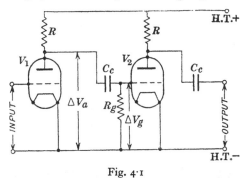

Fig. 4·1

anode circuit resistance, R_g the grid-leak resistance. R_a is the internal slope resistance of the valve. ΔV_a represents a change of anode potential, ΔV_g a change of grid potential. Suffixes 1 and 2 will be used to denote successive valves. It may be seen that

$$\Delta V_{g2} = \Delta V_{a1} \frac{R_g}{R_g + \dfrac{1}{j\omega C_c}}$$

or

$$|\Delta V_{g2}| = |\Delta V_{a1}| \frac{1}{\sqrt{1 + \dfrac{1}{\omega^2 T^2}}},$$

where $T = C_c R_g$. From Fig. 4·2 it is apparent that ΔV_{g2} becomes appreciably less than ΔV_{a1} at frequencies for which $\omega < 1/T$. The amplification therefore falls at low frequencies.

In practice, the amplification also falls at high frequencies, due to the interelectrode capacities of the valves and stray capacities in the circuit.

The effective input capacity of an amplifying valve may be very different from that when the cathode is cold. This arises because a change ΔV_g will produce a change of anode potential $\Delta V_a = -m\Delta V_g$, where m is the magnification of the stage, and

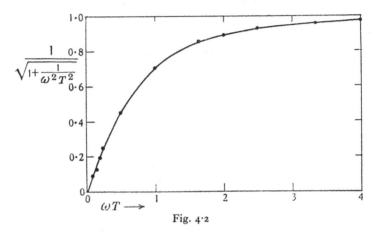

Fig. 4·2

the negative sign indicates that the anode potential falls when the grid potential rises. The resultant change of potential on the grid-anode capacity C_{ga} is then $(1+m)\Delta V_g$; the charge required to produce this change is $(1+m)\Delta V_g C_{ga}$. The effective grid-anode capacity as far as the grid circuit is concerned is thus seen to be $(1+m)C_{ga}$, for a change of grid potential ΔV_g requires a charge $(1+m)\Delta V_g C_{ga}$ (Fig. 4·3). The effective input capacity of the valve is therefore $C_I = (1+m)C_{ga}+C_{gf}$, where C_{gf} is the grid-filament capacity.

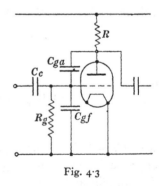

Fig. 4·3

At high frequencies, for which the effect of R_g may be neglected, we have $\Delta V_{g2} = \Delta V_{a1}\dfrac{C_c}{C_c+C_I}$ (Fig. 4·4), but generally $C_c \gg C_I$, and the effective circuit reduces to Fig. 4·5. The anode circuit

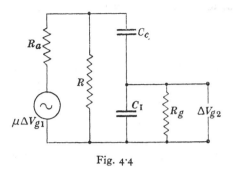

Fig. 4·4

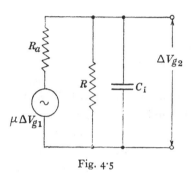

Fig. 4·5

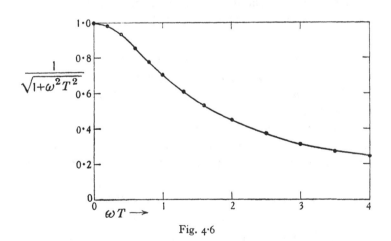

Fig. 4·6

load reduces to R paralleled by C_I, i.e. $\dfrac{R}{1+j\omega C_I R}$, so that the

stage magnification becomes $m = \mu \dfrac{R}{R_a(1+j\omega C_I R)+R}$, which is complex. This affects the input capacity C_I of the first valve. The magnification, neglecting the phase change, is

$$\frac{\mu R}{R+R_a} \frac{1}{\sqrt{(1+\omega^2 C_I^2 r^2)}}, \quad \text{where} \quad r = \frac{RR_a}{R+R_a}.$$

This is illustrated in Fig. 4·6. As has been seen, however, C_I is not constant but is given by C_0+mC_{ga}; the effect of this is to raise the curve somewhat at high frequencies.

Tetrode or triode.

Screen-grid tetrodes or pentodes offer some advantages when compared with triodes for resistance-capacity coupled amplifiers. These screen-grid valves have an extremely small C_{ga}, so that C_I is much smaller and m may be very high. For example, R may be 100,000 ohms and the mutual conductance μ/R_a may be 4 mA./V., so that an amplification of 400 per stage may be obtained if $R_a \gg R$. C_I may be about 15 $\mu\mu$F., to which must be added the anode to screen-grid capacity of the previous valve, about 10 $\mu\mu$F., and the stray capacity of the coupling condenser to earth, about 5 $\mu\mu$F., making a total of 30 $\mu\mu$F. Hence

$$C_I r \simeq C_I R = 3 \times 10^{-6} \text{ sec.,}$$

and amplification will be maintained up to 50,000 c./sec.

If, however, R were reduced to 10,000 ohms the amplification would be 40 times and would be maintained up to frequencies of about 500,000 c./sec.

These figures may be compared with those for triodes. A valve such as the Osram MH 41 may have $\mu = 75$, $R_a = 12,000$ ohms. With an anode resistance of 50,000 ohms the stage gain at medium frequencies would be about 60. C_{ga} is about 4 $\mu\mu$F., so that $C_I = 61 \times 4 +$ incidental stray capacities $= 260 \mu\mu$F.,

$$C_I r \simeq C_I R_a = 2 \cdot 6 \times 10^{-6} \text{ sec.,}$$

and amplification is maintained up to about 60,000 c./sec.

It is thus evident that the tetrode may have an amplification comparable with that of a triode over a wider range of frequencies. Against this apparent superiority of the tetrode it must be remembered that special construction of the amplifier must be adopted with the screen-grid valve to avoid any external increase of C_{ga}, for it is upon the low value of this (about $0 \cdot 002 \, \mu\mu$F.) that its superiority depends. Also in such a high-gain amplifier as is considered it would be necessary to adopt separate decoupling

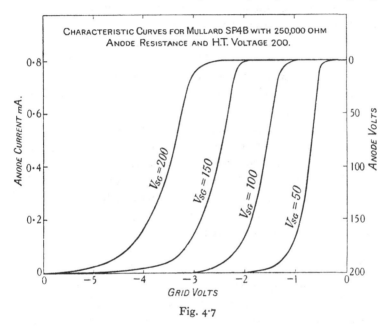

CHARACTERISTIC CURVES FOR MULLARD SP4B WITH 250,000 OHM ANODE RESISTANCE AND H.T. VOLTAGE 200.

$V_{SG} = 200$
$V_{SG} = 150$
$V_{SG} = 100$
$V_{SG} = 50$

ANODE CURRENT mA.

ANODE VOLTS

GRID VOLTS

Fig. 4·7

for the screen grids. In view of these complications, and also because triodes have proved perfectly satisfactory for the present purpose, the screen-grid valve is not always adopted. In America it is more commonly used, as the American screen-grid valves have mutual conductances of about $1 \cdot 5$ mA./V., and the practice has been to keep to a stage gain of not more than 100, and to use the same number of stages as in the equivalent triode amplifier.

To obtain a satisfactory performance from an amplifier using screen-grid valves the grid-bias voltage must be kept within narrow limits. This is well illustrated by the family of curves (Fig. 4·7) for a screen-grid pentode (the Mullard SP 4 B) with an anode circuit resistance of 250,000 ohms and an anode supply voltage 200. The amplification falls when the anode voltage becomes very low, and also when the anode current becomes very low. For a given screen-grid voltage the grid-bias voltage at which the anode current is cut off is almost independent of the anode voltage. Moreover, if the valve provides an amplification of 400 and has an anode voltage of 200 a change of grid bias of 0·5 V. would cover the whole working range. This indicates the limits within which the grid bias must be fixed. It is often convenient to arrange that the grid bias is controlled automatically by the average cathode, anode or screen-grid currents. Control by the cathode current is probably most commonly adopted. This is illustrated in Fig. 4·10 (*e*), (*f*).

It is useful to note that the ratio of screen-grid to anode current is almost constant for any particular valve. If the screen grid is connected to the anode it is easily seen that the cathode current will divide between the two in a certain ratio depending on the geometry of the electrode structure. Moreover, the anode current is almost independent of the anode voltage over the working range, so this ratio is preserved.

Design of the ionization pulse amplifier.

Five magnifying stages are generally more than sufficient to give all the amplification that can usefully be employed. With a special design fewer stages could be made to give the same full output, but this practice is not generally adopted. Only three stages of the normal amplifier are used to give a very high voltage gain. Amplification is sacrificed in the first stage in order to secure as low a ratio of background to pulse as possible. The final stage is generally designed to give the maximum output power and not the maximum voltage gain.

The overall voltage amplification required is between 10^6 and 10^7 (i.e. 120–140 db.). Both with triodes and with tetrodes it is easy to exceed this amplification with the five stages. If the first and last stages give an amplification of 10 and the other three stages a gain of 60 each, the overall amplification would be $2 \cdot 16 \times 10^7$. Usually such a high amplification is not required, and it is possible to secure a greater constancy of amplification by adopting negative feed-back to remove the excess amplification; this important principle is discussed in Chapter VI.

It is not required that an ionization-pulse amplifier should have uniform amplification over a wide frequency range, but since the response of any linear amplifier to a transient impulse is a unique function of its amplification-frequency characteristic, the form of this characteristic is very important. Consider how the pulse amplifier is required to act. A positive charge is collected in a time which may be from 10^{-2} to 10^{-4} sec. on an electrode connected to the control grid of the first valve. This charge leaks away relatively slowly, the discharge time constant being between 1 and 10^{-3} sec. This impulse must be registered, and perhaps measured, in a time as short as possible so that the recorder is left free to register another impulse with a minimum of delay. If Fig. 4·8 (*a*) represents the voltage change on the first grid, (*b*), (*c*), (*d*) represent the impulse after passing through one, two and three coupling stages of the amplifier. If the time constants $C_c R_g$ of these stages were made smaller, the ripple following the impulse would be larger in amplitude and shorter in time. As the base-line must be preserved for measuring subsequent impulses, the ripple should be kept small in amplitude, so that $C_c R_g$ must clearly be as large as possible. But even then a large amount of time is wasted on the downslope of the impulse. Most of this time may be recovered for use by letting one and only one of the coupling stages have a small $C_c R_g$. The result of this is shown at *e*.

This may be regarded in two ways. By including one quick stage the amplification at low frequencies is reduced so that the steep upslope is still amplified but the slower downslope is less amplified. Alternatively, we may say that by having a small $C_c R_g$ the voltage cannot be maintained on one side of the axis for a time much longer than $C_c R_g$.

The reason why only one quick stage or short time-constant

coupling may be introduced may be noted. If two were introduced the back pulse after the impulse would be comparable in magnitude with the impulse itself. If more than two quick stages were introduced the ripple after the impulse would also be large so that every impulse would appear at least double. It is seen that

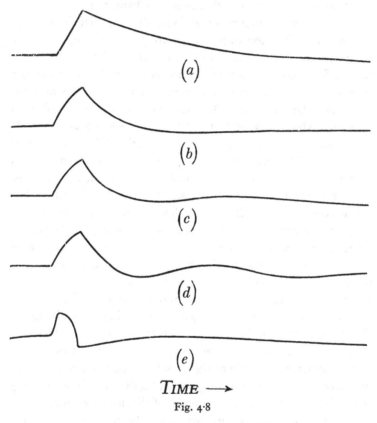

Fig. 4·8

the quick stage is only of advantage when the time of collection of ions is short compared with the recovery time of the collecting electrode and the grid of the first valve. If this recovery time cannot be made long it is sometimes an advantage deliberately to make it short and to use no other quick stage in the amplifier. This procedure is, however, not generally to be recommended owing to its effect on the background. Most of the background originates inside the first valve, and if all the coupling stages following this valve have long time constants the low frequencies in the background, which are usually large due to flicker effect and imperfect grid insulation, are fully amplified, with the result that the grid voltage of the penultimate or of the output valve may be varying over such a wide range at low frequency that the amplification of short impulses may not be constant. For this reason also it is an advantage that the quick coupling stage should be one of the early coupling stages in the amplifier.

As a typical example of the orders of magnitude of impulses at the various stages of the amplifier (without negative feedback) it may be taken that the original impulse on the grid of the first valve is $+ 100 \mu$V. On the grids of the second, third and fourth valves it may be $- 1$ mV., $+ 40$ mV., $- 1 \cdot 6$ V. respectively. The fourth valve, the penultimate valve, is usually followed by some potentiometer device for controlling the size of the impulse passed to the output stage. The amplification obtainable is conveniently such that $+ 10 \mu$V. on the first grid can produce $+ 10$ V. on the grid of the output valve, but for normal operation the amplification is reduced below this level.

It is necessary to decide how the three supplies for anode filament and grid bias are to be obtained. There is little difficulty about the high-tension anode supply; whether a battery or rectified a.c. is used it will be necessary to decouple the supply to each stage in the well-known manner (Fig. 4·9), with decoupling condenser C_d and resistance R_d. The time constant $C_d R_d$ should be long compared with the coupling time constant $C_c R_g$. If rectified a.c. is used, it is usually desirable that the voltage should be stabilized in some manner to prevent changes of amplification with changes in the mains voltage. This may most simply be done by neon lamp stabilizers (see Fig. 6·8), or more effectively by a valve stabilizer circuit such as described in Chapter VI. If very low frequencies are to be amplified, so that it would be expensive to provide adequate decoupling, the high-tension supply may be separated into two parts, one for the earlier stages and one for the later stages of the amplifier.

When the high-tension supply is derived from rectified a.c. it must be adequately smoothed. It is necessary for the smoothing to be more complete for an amplifier using screen-grid valves than for one using triodes, for if a ripple voltage $\varDelta V$ is introduced in series with the anode resistance, the voltage appearing on the grid of the following valve is proportional to $\dfrac{R_a}{R+R_a}\varDelta V$; this

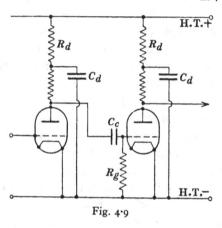

Fig. 4·9

voltage is therefore greater for the screen-grid valve for which R_a is large.

The question of filament supply has no such unique solution. Undoubtedly the simplest is to use accumulators for heating at least the first valves of the amplifier. It is possible to heat all the valves with a.c. if the wiring is very carefully laid out, and the transformer is designed with balanced secondary windings, the exact mid-point of which is earthed. For a.c. heating the cathodes of the earlier valves must be of a well-designed indirectly heated type. Reliable quantitative information of the minimum hum level practically obtainable does not appear to be available. Alternatively, valves having indirectly heated cathodes may be supplied with rectified a.c. perfectly satisfactorily. The writer has experience of one such amplifier in which the filaments of

the first two valves are heated in this way. The valves used take
o·3 amp. at 8–13 V. and are run in series, the current being
regulated by a baretter from a supply at 150 V. obtained from
rectified a.c. and generously smoothed. This method is extra-
vagant but safe; valves may be changed with confidence that hum

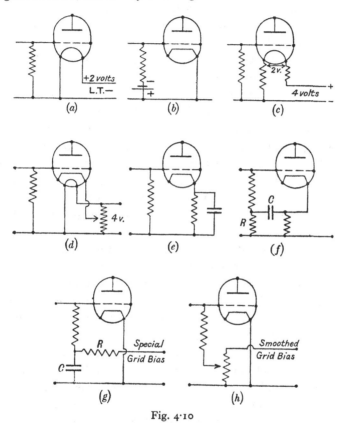

Fig. 4·10

will not be introduced. The heaters of indirectly heated cathode
valves used in later stages of the amplifier may be heated by a.c.
without difficulty.

Many methods, none of them ideal, may be adopted for obtaining
grid bias. Some of these are illustrated in Fig. 4·10. If the early
valves of the amplifier are 2 V. battery valves of the H type, grid

bias is unnecessary if the negative end of the filament is earthed as shown in Fig. 4·10(*a*). The only valve on which the input grid swing is of the order of 1 V. receives negative impulses so that grid current is avoided. Grid bias is only required for the output valve. An extremely simple amplifier is possible using battery valves of this H type with an accumulator for filament heating; such an amplifier is described in more detail later. The provision of grid-bias cells for individual valves (shown in Fig. 4·10(*b*)) is not to be recommended, though it ensures good stability of amplification. The disadvantage is that when the amplifier produces, as every amplifier does at some time or another, an increased and irregular background the grid-bias cells may be responsible, and in the design of practical amplifiers the number of such possible sources of trouble must be kept to a minimum.

When an accumulator is used for heating the first valves grid bias may often be derived from this if the voltage is greater than required for heating the valves. This is shown in Fig. 4·10(*c*). Often the first valve may require a 4 V. accumulator and the later valves only 2 V. Fig. 4·10(*d*) shows a method applicable with indirectly heated valves where an accumulator is used, grid bias being obtained from potentiometers across this accumulator. Again, with indirectly heated valves bias may be obtained as in Fig. 4·10(*e*), (*f*) from the cathode current flowing through a resistor between the cathode and earth. In radio- and audio-frequency amplifiers it is common to shunt such self-bias resistors by large-capacity condensers. It is expensive, however, to make the time constant of this combination sufficiently long, and such electrolytic condensers are not sufficiently reliable to be included in this position in amplifiers of the type under discussion. It is better therefore to use no condenser across this bias resistor. The feedback at low frequencies due to the bias resistor will be negative, with the effect that the amplification of the stage is reduced (see p. 47).

Fig. 4·10(*f*) shows another method of applying grid bias. In this *CR* is made large compared with the lowest frequencies to be amplified. This is unsatisfactory if grid current should happen

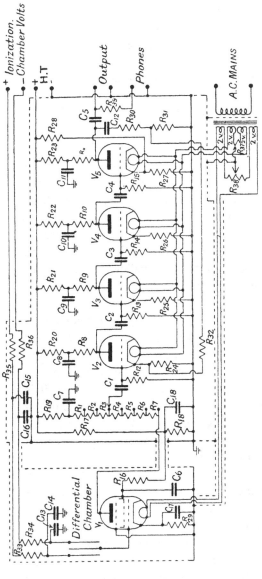

Fig. 4·11. Stabilized high-gain amplifier with a.c. filament heating.

RESISTANCES

Anode circuits: R_1, 20,000 ω; R_2, R_3, R_4, R_5, 50,000 ω; R_8, R_9, R_{10}, R_{11}, 100,000 ω.

Grid circuits: R_{12}, 125,000 ω; R_{13}, R_{14}, 2 MΩ; R_{15}, 250,000 ω.

H.T. supply: R_{16}, R_{17}, R_{18}, 250,000 ω; R_{19}, 100,000 ω, R_{20}, R_{21}, R_{22}, 50,000 ω; R_{23}, 20,000 ω.

Cathode circuits: R_{24}, R_{25}, R_{26}, R_{27}, 1000 ω; R_{28}, 100,000 ω; R_{29}, 250,000 ω.

Feed-back circuits: R_{30}, R_{32}, 1 MΩ; R_{31}, 20,000 ω.

Chamber volts: R_{33}, R_{34}, 2 MΩ; R_{35}, R_{36}, 5 MΩ.

Hum balancers: R_{37}, 0·12 ω; R_{38}, 0·6 ω (bare resistance wire).

Phone limiter: R_{39}, 100,000 ω.

CONDENSERS

Coupling: C_1, 0·005 μF.; C_2, C_3, 0·01 μF.; C_4, 0·004 μF.; C_5, 0·1 μF.

Decoupling: C_6, 0·1 μF.; C_7, C_8, C_9, C_{10}, C_{11}, 1 μF.

Feed-back: C_{12}, 0·1 μF.

Chamber: C_{13}, C_{14}, C_{15}, C_{16}, 0·05 μF.; C_{18}, 0·5 μF.

Cathode: C_{17}, 0·1 μF.

Valves: V_1, Mullard SP4B; V_2, V_3, V_4, V_5, Mullard 904 V.

Transformer: All-power transformers, Ltd., 8 a Gladstone Road, Wimbledon, S.W. 19. Each pair of secondaries symmetrically placed on either side of the screened primary winding.

3·2

to flow momentarily during an abnormally large impulse, for then the grid bias is increased until the charge has leaked off the capacity C. A further modification is shown in Fig. 4·10(g), which is liable to the same grid-choking effect. The general wiring of the amplifier is also complicated by these systems.

Grid bias for the output stage may be obtained by the use of a bias battery (Fig. 4·10(b)) or by some method such as Fig. 4·10(f), (g), (h).

It remains to discuss the layout of the amplifier. It is almost always essential that the amplifier should be screened as completely as possible against stray electromagnetic fields in the laboratory. It should therefore be completely enclosed in a metal box, common tinplate being as good as anything. Further, the amplification is so great that the input wiring of the amplifier must be carefully screened from the output wiring.

The very simplest form of construction that can be adopted is a long flat box, with the input at one end of the length and the output at the other. By keeping the box shallow the capacity of all the components is mainly to the box and interaction between components is a minimum. The amplifier shown in Figs. 4·11 and 4·12 is made on this principle, which is highly recommended where the circuit is without complications such as elaborate grid-bias circuits, gain control potentiometers and bulky coupling and decoupling condensers, which are required when very low frequencies have to be amplified. Where this is required it is perhaps better to depart from the two-dimensional layout and construct a box with screening partitions. One such design is shown in Figs. 4·13 and 4·14. Where screen-grid valves are used a similar design must be adopted.

When planning the wiring it should be borne in mind that such resistance-capacity coupled amplifiers are prone to three kinds of parasitic oscillation. The simplest is a low-frequency oscillation known, from the sound produced in telephones, as motor-boating. This is always due to insufficient decoupling in the high-tension circuits or in the grid-bias circuits if that of Fig. 4·10(g) is used. Motor-boating is unlikely to occur if the design

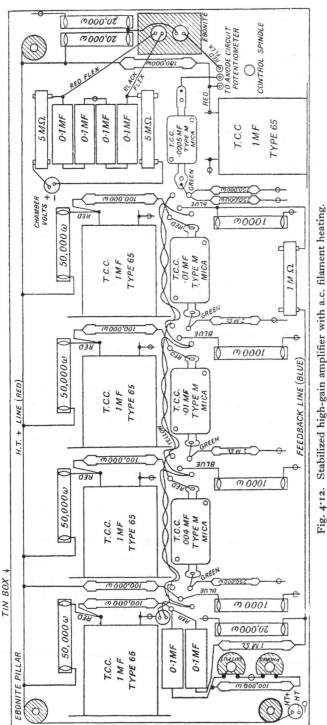

Fig. 4·12. Stabilized high-gain amplifier with a.c. filament heating.

RESISTANCES

1000 ω, 20,000 ω, 50,000 ω, Watmel "Hywatt" wire-wound.
100,000 ω, 1 MΩ, Siemens-Schuckert "Karbowid" Type 3B.
250,000 ω, 2 MΩ, Siemens-Schuckert "Karbowid" Type 2B.
1 MΩ, 5 MΩ, Siemens-Schuckert "Karbowid" Type 4A.
Phone: Jacks Igranic Midget.
Condensers: 0·1 MF T.C.C. Type 341.

KEY TO SYMBOLS

Wire soldered to tin ——⊙——
Wires joined ——•——
Wires crossing ——+——

has been thought out on the lines indicated. On the other hand, a type of oscillation difficult to distinguish by ear from motor-boating is liable to be encountered. This consists of an oscillation at a frequency above the audible range, which, due to the flow of grid current, charges one of the grids so negative that a valve ceases to amplify, thus removing the feedback which was maintaining the oscillation. When the grid discharges, the oscillation begins again and the process is cyclically repeated at a low frequency. This phenomenon is known as an "automatically interrupted" "blocking" or "squegging" oscillation. To cure it the feedback causing the supersonic oscillation must be removed. Such a supersonic oscillation is not necessarily interrupted. It may readily be produced by taking off part of the screening near the input end of the amplifier and pushing in through the opening a wire connected to the output. As the wire is approached closer the frequency is lowered and becomes audible. The feedback in this instance is capacitive, and it should be particularly noted that this never fails to produce an oscillation, there being always some frequency at which the feedback is in the correct phase. As the feedback capacity is made larger the frequency at which this phase relation is correct naturally moves to a lower frequency. Such feedback can accidentally occur in the wiring of the amplifier due to the common supply wires, particularly the filament and grid-bias wiring. Since the oscillation is at a high frequency it is often possible to cure it by connecting a small 0·01 or 0·1 μF. condenser between the responsible wire and the screening.

The third type of oscillation which may be encountered occurs at a frequency of about 50 or 100 Mc./sec. (6 to 3 metres wavelength). The circuit shown in Fig. 4·15 is one of the most effective for generating such frequencies, and it will be seen that such a circuit might unintentionally occur in the wiring of a resistance-capacity coupled amplifier. The oscillatory circuit is formed by the wires A and B, the stray capacity C, and the valve interelectrode capacity. If such an oscillation occurs it should be possible by moving the wire A, or touching it at different points,

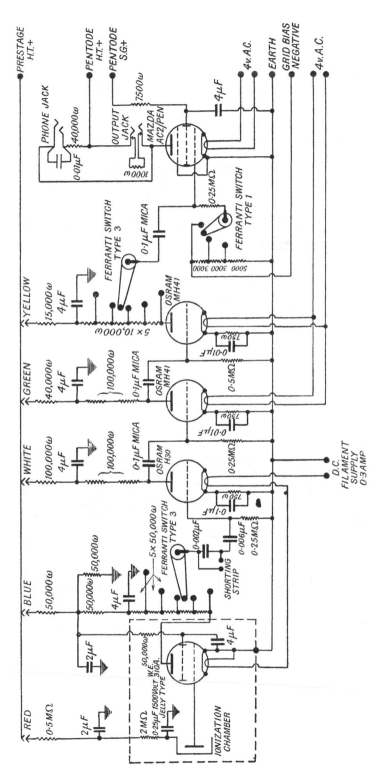

Fig. 4·13. Circuit diagram of amplifier of Fig. 4·14.

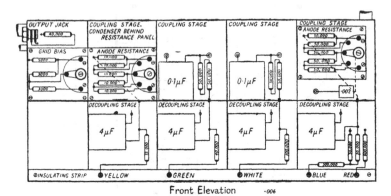

Front Elevation   ·006

- ⊸ Wire soldered to tinplate.
- • Junctions.
- +— Tinplate screening partitions.
- ◎ Insulating bush through tinplate to back.
- ⊙ Plug and socket to H.T.+ busbar.
- ■ Insulating bush through tinplate.

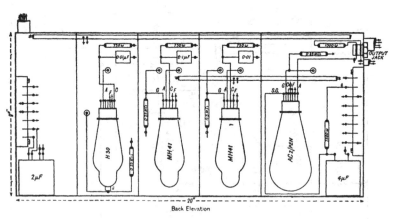

Back Elevation

Fig. 4·14. Screened amplifier.

to alter the amplitude of the oscillation and hence the anode current of the valve. If the wires *A* and *B* are very short (not more than an inch from the valve pins) and the stray capacity *C* is small, such an oscillation should not be encountered. If it is inconvenient to alter the disposition of the wiring the oscillation may be stopped by making the wires *A* and *B* of thin-resistance wire or by including a resistance of 100 ohms of small self-capacity in series with one of the wires close to the valve pins.

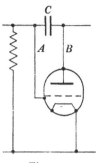

Fig. 4·15

In wiring an amplifier it should be remembered that an electrical contact is satisfactory if the *contact pressure* (force *per unit area*) is sufficient. The contact area must also be sufficient to avoid local heating which may lead to oxidation at the contact, that is, it must be sufficient for the current to be carried. It is quite possible to use plug-in valve sockets throughout the amplifier, but this is not recommended. The requirements of sufficient contact pressure and contact area are likely to fail to be satisfied first for the filament contacts. In practice every amplifier at some time gives rise to a bad background; when this occurs every spring contact carrying current is to be suspected as a possible cause of the trouble. For this reason spring contacts carrying current should be avoided where possible; soldered joints are recommended.

The output stage.

The design of the output stage depends on the method adopted for measuring and recording the impulses. If a record is to be taken by photographing the trace of an electromechanical oscillograph, the output stage should be capable of delivering 1 or 2 W. at any frequency in the range of the amplifier. If merely a valve or thyratron counter is to be operated, much less power is required from the output stage. Whether the powerful output stage should be regarded as part of the oscillograph equipment or part of the amplifier equipment depends on circumstances. An output stage capable of operating a mechanical oscillograph will also be suitable for operating a valve or thyratron counter.

A pentode output valve is preferable to a triode for operating an oscillograph, since the reactance of the oscillograph is mainly inductive at frequencies to which it responds. The internal anode-slope-resistance of a pentode is much higher than that of a triode, so that the time constant L/R of the oscillograph in series with the valve is much smaller for the pentode than for the triode, and the response of the oscillograph is therefore better maintained at high frequencies when a pentode is used.

Chapter V

OSCILLOGRAPH RECORDING

It is often desirable to have a permanent record of the impulses from an amplifier. Photographic records from an electro-mechanical oscillograph have been found very suitable.

The aim in the design of an electromechanical oscillograph is to produce a mechanical movement which is an accurate repro-duction of an electrical wave form. One type of oscillograph is a form of galvanometer which has its lowest natural frequency higher than the highest component frequency in the wave form to be reproduced.

If the mechanical equation of motion of the moving part of the oscillograph is $\theta'' + 2k\theta' + \omega^2\theta = G/I$, where k includes the electromagnetic damping resulting from the reaction of the moving part on the electrical circuit. I is the moment of inertia about the axis of rotation and θ is the angular displacement about the same axis; G is the torque acting due to the current in the electrical circuit. If G is sinusoidal $= G_0 \cos pt$, then

$$|\theta| = \frac{G_0}{I\sqrt{[4p^2k^2 + (p^2 - \omega^2)^2]}}.$$

Then as long as $p^2 \ll \omega^2$ and $4k^2 < \omega^2$ the amplitude of θ, and ψ, the relative phase of θ and G, $\left[\tan\psi = \frac{p^2 - \omega^2}{2pk} \right]$, is independent of frequency. By a suitable choice of k, say $\frac{1}{2}\omega < k < \omega$, these conditions can be maintained up to frequencies p which are a large fraction of ω. Under these conditions θ will be an accurate reproduction of G and, since the equation of motion is linear, the same is true for any form of G provided it does not contain components of higher frequency.

It should perhaps be noted that between $p = 0$ and $p = \omega$ the relative phase of θ and G necessarily changes by $\frac{1}{2}\pi$, but this does not imply that θ is necessarily a distorted reproduction of G. If the relative phase changes linearly with frequency, the form of θ is an accurate reproduction of G but delayed in time.

In practice it is not difficult to make the frequency $\omega/2\pi$ as high as 2000 c./sec., while retaining adequate sensitivity for operation from an ordinary valve of 6 W. dissipation. The moment of inertia I cannot be made very small, since it is necessary for the moving element to carry a mirror of adequate size for the photographic registration of its quickest movement. The brightness of the mirror cannot exceed the brightness of the light

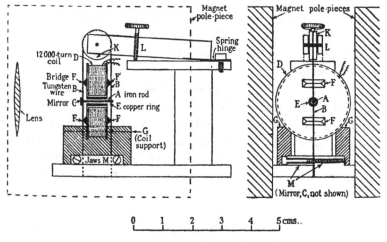

Fig. 5·1. Wynn-Williams oscillograph.

source. A very intense source, such as a carbon arc or a high-pressure mercury arc, is therefore desirable.

An excellent oscillograph of this type has been described by Wynn-Williams (81) (Fig. 5·1). The armature consists of a small soft iron rod suspended by taut tungsten wires so as to lie along the axis of the fixed oscillograph coil. The whole is placed between the poles of a strong magnet giving a field of about 2000 gauss. The armature thus occupies a position similar to that of the armature in the balanced armature type of loud-speaker movement, the main magnetic field being at right angles to the length of the armature. A small mirror is attached to one end of the armature. The tungsten wires are tensioned so that the lowest natural frequency of the system is about 3000 c./sec. The sensitivity of this type of oscillograph is about 1 cm. deflexion at 1 m. for 5 mA. in the coil which has an impedance of about 15,000 ohms at 1000 c./sec. In order that the time constant of the electrical circuit shall be low, the oscillograph is operated in the anode circuit of a pentode having a high-anode slope resistance.

An oscillograph of greater sensitivity, but rather lower natural frequency, is on the market, being made by Muirhead.

A quite distinct method of attacking the problem is first to differentiate the wave form with respect to time and then record with an integrating device. For example, a transformer may be used to perform the differentiation and a fluxmeter of a type having a negligible restoring torque may be used to integrate the resultant wave form. A very satisfactory oscillograph operating on this principle has been designed by Shire (67) (Fig. 5·2). The moving system is a light aluminium single-turn loop. It is suspended by a torsionless suspension so as to be free to rotate about an axis of symmetry in its plane. It is looped round the laminated core of the transformer and lies between the poles of a powerful magnet which produces a field at right angles to the transformer core. Virtually the same principle has been applied in the type of moving-coil loud-speaker movement known as the Duode. The performance of this oscillograph is indicated by the characteristics of one instrument, which in a field of 12,000 gauss reached 95 % of its final deflexion in 1/5000 sec. after applying a voltage suddenly. Its sensitivity was 1 cm. deflexion at 1 m. for 4 mA. in the primary which had an inductance less than 1 H.

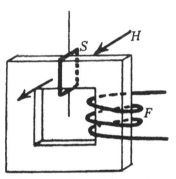

Fig. 5·2. Shire oscillograph.

Where it is necessary to delineate a wave form containing very high frequencies the cathode-ray oscillograph must be used.

A great deal has been written on the technique of operating cathode-ray oscillographs, so this will not be discussed here. It should, however, be remarked that the use of a small cathode-ray oscillograph with a linear time base traversed in about 1/15 sec. is very valuable for the visual monitoring of the output from an impulse amplifier. A larger cathode-ray equipment providing a linear horizontal time sweep with adjustable timing from about 1/15 to 1/10,000 sec., adjustable gain on the amplifier controlling the vertical deflexion, and facilities for controlling the width of the horizontal sweep, is also found invaluable for designing circuits and setting up counting installations with any special properties.

Chapter VI

FEEDBACK AND STABILIZERS

Owing to the fact that general practice in the application of valve circuits at present falls far short of what is known to be possible, in this chapter certain principles which are likely to become more widely applied will be described.

The principle of feedback and in particular of negative feedback [8] is a most powerful tool in the hands of the circuit designer. Consider the voltage amplifier circuit shown in Fig. 6·1. An input

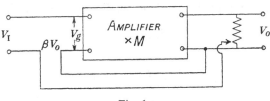

Fig. 6·1

voltage V_I gives rise to an output voltage V_O. A fraction of this βV_O is fed back to the input of the amplifier. The voltage V_g between the input terminals of the amplifier is therefore $V_g = V_I - \beta V_O$. If M be the voltage magnification produced by the amplifier, then $V_O = M(V_I - \beta V_O)$, from which we obtain

$$V_O = MV_I/(1 + M\beta). \qquad (6·1)$$

We note in particular that if $M\beta \gg 1$ then $V_O = \frac{1}{\beta} V_I$ and is independent of the voltage amplification of the amplifier provided this is sufficiently great. Now it is easy to arrange in practice that β is a constant quantity, so that the amplification is constant and independent of changes in the amplifier itself. Take, for example, an amplifier which has four stages having a gain of 40 per stage giving a possible amplification M of 2,560,000 times, and suppose that an amplification one-twentieth of this is all that is required.

If a fraction $\beta = \dfrac{1}{128,000}$ of the output is fed back in the correct phase into the input circuit the amplification will be

$$\frac{1}{\beta}\frac{1}{1+1/M\beta} = \frac{128,000}{1\cdot05} = 122,000,$$

and if M changes by 20 % the amplification changes by only 1 %.

Negative feedback has a pronounced effect on the internal resistance of the output circuit. We have

$$V_O = M'V_g\frac{Z}{R_a+Z}, \tag{6·2}$$

where V_g is the voltage between the input terminals of the amplifier, Z = output load, R_a = effective internal resistance of the output stage of the amplifier for $Z \gg R_a$. Substituting $\dfrac{M'Z}{R_a+Z}$ for M in equation (6·1) leads to

$$V_O = \frac{M'V_I Z}{R_a+Z}\frac{1}{1+\beta M'Z/(R_a+Z)}$$

$$= \frac{M'V_I}{1+\beta M'}\frac{Z}{R_a/(1+\beta M')+Z}.$$

Comparing this with equation (6·2) it appears that the effective internal resistance of the output circuit is reduced by the feedback to $\dfrac{R_a}{1+\beta M'}$. This is applied, for example, in the circuit of Fig. 6·2, which is called the cathode follower circuit because the potential of the cathode follows very closely the potential of the control grid (10, 46). It will be noticed that $\beta = 1$, the effective internal resistance of the valve is therefore $\dfrac{R_a}{1+\mu}$, where μ is

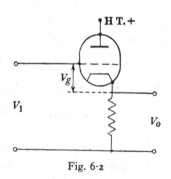

Fig. 6·2

the amplification factor. For example, if the valve has

$$R_a = 2400 \text{ ohms}$$

and $\mu = 15$ the effective internal resistance is 150 ohms. It will be noted that this effective internal resistance is approximately the reciprocal of the mutual conductance μ/R_a. The circuit is extremely valuable since the input impedance is very high and considerable power amplification is obtained, although the output voltage is approximately equal to the input voltage, being given by $V_O = V_I \dfrac{M}{1+M}$. The same principle may be extended to more than one valve, and the circuit of Fig. 6·3 has been found useful

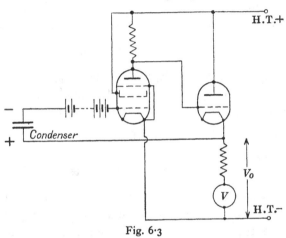

Fig. 6·3

for following the voltage on a condenser from which no current could be drawn. For such a circuit M might well be as great as 400, so that V_O is closely equal to V_I and the effective internal resistance of the output circuit may be a fraction of an ohm. It has been assumed that the high-tension voltage is maintained constant.

Feedback also has an effect on the frequency range over which amplification is maintained. We have seen that at high frequencies the amplification of a single-resistance capacity-coupled stage is $M = \dfrac{\mu R}{R+R_a} \dfrac{1}{\sqrt{(1+\omega^2 C_I^2 r^2)}}$. Write $C_I r = t$ and $M_0 =$ amplification at medium frequencies $= \dfrac{\mu R}{R+R_a}$ for a single stage. Then at high frequencies for which $\omega^2 t^2 \gg 1$ the amplification reduces to $M = M_0/\omega t$. For two stages the amplification is $M = M_0/\omega^2 t_1 t_2$, and for s stages $M = M_0/\omega^s(t_1 t_2 \dots t_s)$, where t_1, t_2, etc. are the time constants $C_1 r$ of the successive stages.

With negative feedback the amplification becomes

$$\frac{M}{1+M\beta} = \frac{M_0}{\omega^s(t_1 t_2 \dots t_s) + M_0 \beta}.$$

Without feedback the amplification is reduced to $\dfrac{1}{n} M_0$ at a frequency for which $\omega^s(t_1 t_2 \dots t_s) = n$. With feedback it is reduced to $1/n$ of the amplification at medium frequencies, i.e. to $\dfrac{1}{n}\dfrac{M_0}{1+M_0\beta}$, when $\omega^s(t_1 t_2 \dots t_s) + M_0\beta = n(1 + M_0\beta)$, i.e. $\omega^s(t_1 t_2 \dots t_s) = n(M_0\beta + 1 + M_0\beta/n)$. When $M_0\beta \gg 1$ and $n \gg 1$, this reduces to $\omega^s(t_1 t_2 \dots t_s) = nM_0\beta$. The amplification is maintained above a given fraction $1/n$ of the maximum up to a frequency $\omega = \sqrt[s]{\dfrac{n}{t_1 t_2 \dots t_s}}$ without feedback and to a frequency $\omega = \sqrt[s]{\dfrac{nM_0\beta}{t_1 t_2 \dots t_s}}$ with feedback. It may be noted that $1/\beta M_0$ is the factor by which the amplification has been reduced by negative feedback. If, for example, we have a two-stage amplifier and $\beta M_0 = 100$, the amplification will remain uniform up to a frequency range 10 times that without feedback. It must be remembered that other properties of the negative feedback amplifier such as the lowered output circuit impedance will not be maintained over the extended frequency range.

The stability of amplifiers controlled by negative feedback requires special consideration. This arises because at both ends of the amplified frequency range the phase changes so that the feedback becomes positive or regenerative, and, to preserve stability, it must be ensured that the amplification falls sufficiently low at frequencies where the feedback phase is reversed.

The particular case of the resistance-capacity coupled multistage amplifier may be considered. For low frequencies the effective coupling element is Fig. 6·4, from which it may be seen that $V = E \dfrac{1}{1 + \dfrac{1}{j\omega CR}}$. Write $\tan\theta = \dfrac{1}{\omega CR}$. Then

$$V = E\frac{1}{1 - j\tan\theta} = \frac{E\cos\theta}{\cos\theta - j\sin\theta} = E\cos\theta e^{j\theta}.$$

So that at each coupling stage the voltage phase is turned through an angle θ depending on the frequency ω and the time constant CR, and its magnitude is reduced by a factor $\cos\theta$.

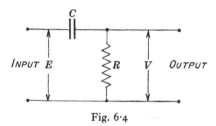

Fig. 6·4

Consider an amplifier having three resistance-capacity couplings. Let the magnification where the phase change is negligible be M. Then the output $V_O = V_g M \cos\theta_1 \cos\theta_2 \cos\theta_3$ and the phase shift is $\theta_1 + \theta_2 + \theta_3$. Suppose a negative feedback voltage βV_O is applied at all frequencies; then if $\theta_1 + \theta_2 + \theta_3 = \pi$ the feedback will be positive and the amplifier will oscillate if $\beta M \cos\theta_1 \cos\theta_2 \cos\theta_3 > 1$. If

$$\theta_1 = \theta_2 = \theta_3 = 60°, \quad \cos\theta_1 \cos\theta_2 \cos\theta_3 = \tfrac{1}{8},$$

so that βM cannot exceed 8 without instability occurring. βM can be much larger, however, if one of the angles, say θ_3, is made larger. Suppose $(CR)_1 = (CR)_2 = 100(CR)_3$. Then when $\theta_1 = \theta_2 = 45°$, $\tan\theta_3 = 100$ so $\theta_3 \simeq 90°$ and $\theta_1 + \theta_2 + \theta_3 = 180°$, but now $\cos\theta_1 \cos\theta_2 \cos\theta_3 = 1/200$, so that βM can have any value less than 200 without oscillation occurring.

It may be recalled that $1/\beta M$ is the factor by which the amplification is reduced by feedback. If M varies by a small amount, the resulting change in amplification of the controlled amplifier is $1/\beta M$ of the variation in M.

High-frequency stability may be treated in a similar manner for a resistance-capacity coupled amplifier using pentodes if the effective grid-anode capacity may be neglected. The effective circuit is Fig. 6·5, from which we see that

$$V = \frac{\mu R}{R+R_a}V_g \frac{1}{1+j\omega C \dfrac{RR_a}{R+R_a}}.$$

Write $\tan\theta = \omega C \dfrac{RR_a}{R+R_a}$. Then $V = \dfrac{\mu R}{R+R_a}V_g \cos\theta e^{-j\theta}$. Hence again at each stage the voltage phase is turned through an angle θ and reduced in magnitude by a factor $\cos\theta$, so the conditions for stability at high frequency are formally similar to those for low-frequency stability.

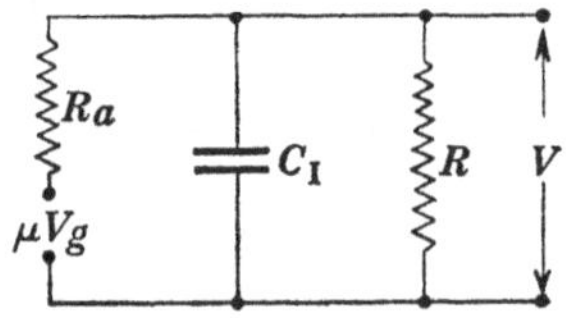

It is generally too complicated to take into account the effect of the grid-anode capacity, and in practice the amplifier is stabilized by adjustment of the input capacity of one stage.

Fig. 6·5

The frequency characteristic of an amplifier may also be controlled by the feedback circuit if this is given a frequency-dependent characteristic. In this connexion there is an important principle which should be recognized, that a unique relation exists between the form of the output impulse from a given input impulse and the frequency characteristic of a *linear* amplifier. If, therefore, by controlled feedback an amplifier is produced with the same frequency characteristic as any other given amplifier, the output from a given input impulse will have the same transient form.

In Chapter IV the form of the output impulse from an amplifier having a coupling stage of short time constant was derived. The form of the amplification characteristic at low frequencies is concerned; this (p. 23) is $M \dfrac{1}{\sqrt{(1+1/\omega^2 T_1^2)}}\dfrac{1}{\sqrt{(1+1/\omega^2 T_2^2)}}\cdots,$ where T_2, T_3, etc. are much larger than T_1, so the essential term is

$\dfrac{1}{\sqrt{(1+1/\omega^2 T_1^2)}}$; this produces the steep downslope on the output impulse, the other terms producing the subsequent slow ripples. When $\omega^2 T_1^2 \ll 1$, the term reduces to ωT_1, and the amplification is thus directly proportional to frequency in the frequency range $\omega^2 T_1^2 \ll 1 < \omega^2 T_2^2$, etc. If two coupling stages had short-time constants T_1, T_2, the amplification would be proportional to $\omega^2 T_1 T_2$. It was pointed out that such an amplifier produces a large reversed pulse following each impulse.

The frequency characteristic of a feedback circuit which would give an amplifier the desired characteristic must therefore be such that $1/\beta \propto \omega$ over a certain frequency range.

There exists, moreover, a definite relation between the amplitude and phase characteristics of an amplifier. This relation is not, however, unique, since any phase change proportional to frequency which amounts simply to a time delay is independent of the amplitude characteristic.

So far only feedback proportional to the output voltage has been considered. If feedback proportional to the output current is applied different characteristics are obtained. When feedback is proportional to the output voltage, this output voltage tends to be maintained irrespective of the load, that is, the output impedance is reduced. On the other hand, when feedback is proportional to the output current, this tends to be maintained irrespective of the load, so the output impedance is increased. This may be useful for mixing, for the outputs of such amplifiers may be connected in parallel with a minimum of mutual interaction. A combination of both types of feedback may be employed to give an output impedance unaffected by feedback, and feedback independent of the load; this has some advantages for achieving stability against oscillation.

Stabilizers.

For many purposes it is required that the voltage of a rectified supply should be independent of variations of the load or of the a.c. supply. Sufficient stability may often be obtained from quite simple devices; a few of these will be described.

For the operation of Geiger-Müller counters and similar discharge devices a potential of the order of 1000–2000 V. is required which is constant within a few volts. The current taken by the counter is negligible, so that a device which maintains a constant current through a fixed high resistance is satisfactory.

The simplest in principle is probably that described by Street and Johnson (71) (Fig. 6·6). The potential drop across the resistance R due to the anode current flowing through it is balanced

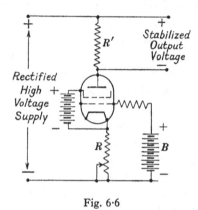

Fig. 6·6

against the voltage of the battery B which may be about 60–100 V. The grid of the valve is thus maintained at a small voltage negative with respect to the cathode. Since the screen-grid voltage is fixed, the anode current is almost independent of the anode voltage over the working range, and is determined solely by the control-grid voltage. Any slight variation in the anode current alters the voltage across R, the resultant change in control-grid voltage tending to minimize the change of anode current. The change of anode current Δi produced by a change of the supply voltage ΔV may be calculated as follows. The change of grid voltage $\Delta V_g = R\Delta i$. The change of anode current is therefore

$$\Delta i = \frac{\Delta V - \mu \Delta V_g}{R + R_a + R'} \quad \text{or} \quad \Delta i(R + R_a + R' + \mu R) = \Delta V. \quad \text{The change}$$

of the stabilized voltage is $\Delta i R' = \Delta V \dfrac{R'}{R'+R_a+(\mu+1)R}$. If, for
example, the stabilized current is 1 mA. and the voltage 1000,
$R' = 1\ \mathrm{M}\Omega$, R_a may be 0·5 $\mathrm{M}\Omega$, $R = 0·1\ \mathrm{M}\Omega$ and $\mu = 1000$. The
change of stabilized voltage is then $\Delta V \dfrac{1}{1·6+100} \simeq \dfrac{\Delta V}{100}$. The
anode voltage of the valve must not be less than the screen-grid
voltage and should never greatly exceed the maker's rating. This
usually allows a variation of about 200 V., so that variations of
$\pm$ 100 V. in 1200 or nearly 10 % in the supply will be com-

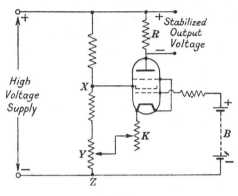

Fig. 6·7

pensated. The circuit has the disadvantage of requiring two high-
tension batteries from one of which a small current is drawn.
The output voltage is conveniently regulated over a small range
by the variable resistance R; larger changes may be made, if
required, by tapping off across only a portion of R'.

A circuit which appears similar but which is fundamentally
different in action is that described by R. D. Evans [21] (Fig. 6·7).
In this the anode and screen-grid currents are maintained exactly
constant by a balance method. The screen grid, control grid and
cathode form a triode, the anode current of which may be written
as $f(\Delta V_{SG}+\mu\Delta V_g)+$ constant, where ΔV_{SG}, ΔV_g are the voltage
changes on the screen grid and control grid and μ is the amplifica-
tion factor of the triode. The ratio of the resistances XY/YZ is

56   *ELECTRICAL COUNTING*

made equal to μ, so that a change of the supply voltage produces changes of screen-grid and control-grid voltages of opposite sign and such that $\Delta V_{SG} = -\mu\Delta V_g$. The triode current and therefore the anode current of the pentode are unchanged. The actual stabilized current through R may be regulated by adjustment of K or of the bias voltage. A slight change of the balance setting will be found for large changes of current or operating voltages. A slight disadvantage of this circuit is that its action must be checked when it is first set up. The tapping point Y is adjusted until the output voltage is independent of variations in the supply voltage. In principle no advantage is gained by increasing the voltage of the battery B, but when the balance adjustment is not quite correct a high-voltage battery B and a large resistance K give increased stability, as the circuit is then similar to that of Street and Johnson.

Stabilized voltages of a lower order are required for high-tension and grid-bias supplies for amplifiers and similar devices. The simplest stabilizer for this purpose is the neon lamp. The circuit is shown in Fig. 6·8. Special tubes are made for the purpose capable of carrying a current of 50 mA. or more (52). The voltage across such a tube (about 100 V.) will not change by more than about 2 V. when the current changes from 10 to

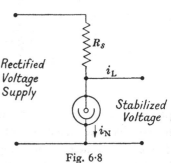

Fig. 6·8

50 mA. From these figures it would appear that the slope resistance of the neon tube is 50 ohms. A tube which has only a small difference between its striking and its burning voltage is advantageous. Suppose the burning voltage is V_b and the striking voltage $V_b + x$. Let V be the supply voltage, R_s the series resistance which includes the effective internal resistance of the supply, and R be the load resistance. Then in order that the neon discharge shall strike

$$V > \frac{(V_b + x)(R + R_s)}{R}, \text{ if the series resistance } R_s \text{ is large, } V \text{ must}$$

be high. If R_s is small, then the current carried by the neon is large and the device is again not economical. There may also be a limit on the total current available. Suppose it is arranged that the neon current i_n is equal to the load current i_l. When the neon discharge strikes, the increase in the voltage drop across R_s is x and this must be less than $i_n R_s$, for the load current will fall somewhat. It follows that, when in operation, the total voltage drop across R_s, viz. $(i_n + i_l) R_s$, is greater than $2x$. Hence the supply voltage V must exceed the burning voltage by at least $2x$, and must be greater if the current carried by the neon stabilizer is to be less than the load current.

Neon stabilizers of this kind may be connected in series as shown in Fig. 6·9. The voltage required to strike the two dis-charges will be $2V_b + x$ if the pilot resistance $R_p \gg R_s$. R_p may be very high, of the order of a megohm. The supply voltage must be greater than $2V_b + 2x$ unless the stabilizer current is greater than the load current. The voltage across R_p is also stabilized, but if an appreciable current is drawn the supply voltage must be increased to ensure that the second stabilizer

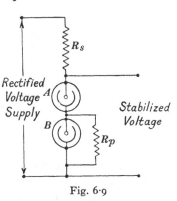

Fig. 6·9

strikes. This defect is minimized if the two discharge gaps are in series in the same bulb; tubes of this kind are on the market.

Small variations of the burning voltage are liable to occur with age and with temperature. The magnitude of these variations is liable to differ between tubes of different manufacture. It is important to allow a sufficient margin for such changes in the design of the stabilizing unit.

Another simple stabilizer with the output voltage adjustable is provided by a neon lamp and a valve connected as a cathode follower (Fig. 6·10). The effective internal resistance of this stabilized supply is $\dfrac{R_a + R_s}{\mu}$, which is rather greater than the reciprocal of the mutual conductance of the valve if R_s is comparable with the anode slope resistance R_a. The output voltage variations should be at most $1/(1+\mu)$ of the supply variations if the valve is worked with a normal anode current and no grid current, and if the internal resistance of the supply is negligible compared with the resistance of the neon-lamp circuit.

Very much better stabilization can be obtained from the use of two valves, and many suitable circuits are available. Two different modes of operation may be distinguished: in the first, variations

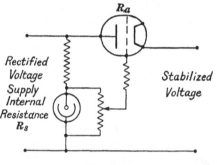

Fig. 6·10

in the output voltage are amplified and applied to the control valve; in the second, variations in the output voltage are balanced out so that control is derived from variations of the supply voltage and of the output current. This second principle might appear at first sight to be preferable, but it is found that while variations of the supply voltage may be compensated over a fair range, variations of load are only compensated over a small range. The effective internal resistance of such a stabilizer may be positive, zero or negative. Against this type of stabilizer it should be noted that the two balances must be made by trial. Also it is difficult or perhaps impossible to design a circuit of this type so that a number of stabilizers operated from one supply have one common output terminal.

The first type of stabilizer may be so efficient, having an

effective internal resistance of only a few ohms, and stabilizing over a wide range of supply and load variations without the necessity for any balancing adjustment, that it is very often to be preferred.

Examples of these two types of circuits are shown in Figs. 6·11 and 6·12. The standard of reference with which the output voltage is compared is provided in these circuits by neon tubes, but batteries may of course be substituted for these.

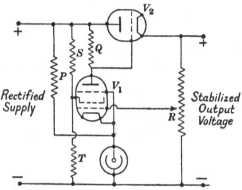

Fig. 6·11. Performance with V_1, Mullard SP4B; V_2, Osram KT41; Neon Mullard 7475; P, 250,000 ω; Q, 250,000 ω; R, 2MΩ; S, 40,000 ω; T, 50,000 ω. Output voltage constant to ± 0·2% for current from 10 to 60 mA. Output current or voltage constant to ± 0·5% for A.C. inputs from 200 to 250 V.

In the circuit of Fig. 6·11 a certain fraction of the output voltage is selected by the potentiometer R and balanced against the voltage across the neon tube in the grid circuit of the pentode. The anode resistance Q of this may be high, so that the amplification of any variation in the output voltage is as great as possible. The pilot resistance P ensures that the neon lamp strikes and passes sufficient current to maintain it in its stabilizing condition. The output voltage may be adjusted by R from the burning voltage of the neon tube up to about 200 V. greater than this provided the supply voltage is sufficient. Typical performance figures are given above, Fig. 6·11.

The grid bias of the amplifying valve V_1 in Fig. 6·12 is obtained from a fraction of the supply voltage selected by the potentiometer P, and a voltage dependent on the load current developed across R. These together constitute a considerable positive voltage which is offset by the voltage across the large cathode resistance. Increase of the supply voltage or decrease of the load current makes the grid potential more positive, thus lowering the anode potential which increases the resistance of the control valve V_2, as is required to maintain the output voltage constant.

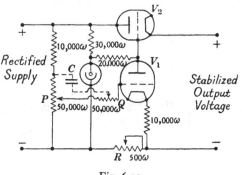

Fig. 6·12

It should be noted that the action of these stabilizers is very rapid, so that they provide a very considerable smoothing of ripple on the rectified supply. In the balanced circuit it may be found that the setting of the potentiometer P for zero hum or ripple is rather greater than that for the elimination of slow fluctuations; this may be compensated by a condenser connected as at C.

The circuit of Fig. 6·11 is almost identical with the valve voltmeter circuit of Fig. 6·3 which was discussed from the viewpoint of negative feedback.

Chapter VII

MIXING CIRCUITS, TRIGGERED CIRCUITS AND DISCRIMINATORS

In the discussion of amplifiers, feedback controlled circuits and stabilizers, valves have been considered as quasi-linear devices. In many other uses the essentially non-linear characteristics of valves are applied.

The mixing circuit of a coincidence counting system is an example of this. A circuit is required such that a pulse is only produced in the output when simultaneous impulses occur in

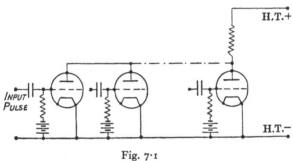

Fig. 7·1

each of a number of independent inputs. This is achieved with the circuit of Fig. 7·1 (57). The anodes of the required number of valves, two for double coincidences, three for triple, four for quadruple and so on, are connected in parallel. The cathodes are similarly connected together but the grids are independent. The anodes are connected to the high-tension supply through a high resistance. The potentials of the grids are such that the internal resistances of the valves (not necessarily the anode slope resistances) are low compared with this high resistance in the anode circuit. Suppose each is a tenth of this, then under normal conditions the common anode potential is $\frac{1}{10}$, $\frac{1}{20}$, $\frac{1}{30}$, etc., of the high-tension potential for 1, 2, 3, etc. valves. Negative impulses are applied to the grids of the valves of sufficient size to stop the flow of anode current even if the anode potential were equal to the high-tension potential. Now as long as one valve has not received such an impulse, the potential of the anodes must remain less than $\frac{1}{10}$ of that of the high-tension supply. If, however, this last valve simultaneously receives such a negative impulse on its grid, the potential of all the anodes rises to the high-tension potential. The change of anode potential when this occurs must be at least nine times that when all but one of the valves received simultaneous negative impulses. This then is a remarkable circuit which passes on a large impulse when all the grids simultaneously receive a negative impulse, but only passes on a much smaller impulse if even all but one of the grids simultaneously receive a negative impulse. Further, the large impulses passed on are all of the same voltage.

Some elaborated circuits have been used for mixing in coincidence systems, but the properties of the simple circuit leave little room for improvement. Although this simple mixing circuit forms the basis of most coincidence counting systems, a complete system may nevertheless be quite elaborate for reasons which are discussed in Chapter XI.

Remembering that the special property of the mixing circuit depends on the non-linear "anode bend" of the valve characteristic, that is to say on the possibility of cutting off all anode current by any voltage on the grid exceeding a certain minimum, it will be realized that the impulses on the grids must be an appreciable fraction of a volt. In order to obtain a good performance with small impulses of this order, it is necessary to choose the type of valve and the operating conditions with care. If large impulses are available, any valve will do. To work with small impulses the valves must have as high a magnification factor μ as possible and must certainly not be of the variable-μ type. It will also be an advantage to work with a high-tension voltage not greater than about μ V. If a higher voltage is used, the impulse applied to the grid will have to exceed about 1 V. if it is to change over the valve from a state in which it passes anode current when the anode voltage is a small fraction of the high-tension voltage, to a state in which it passes no anode current for an anode voltage equal to the high-tension voltage. Certain high-magnification screen-grid or pentode valves may show some advantage. The screen-grid potential is kept fixed and the control grid is biased to within 1 V. of the point at which the screen grid and anode currents are cut off. Under these conditions a suitable valve will pass an appreciable anode current for an anode voltage as low as 5 V., the high-tension voltage may be much greater than the screen-grid voltage so that quite a large impulse is obtainable in the anode circuit. With a good valve the performance is well maintained up to quite high screen-grid and high-tension voltages if the control grid bias is correspondingly increased. The limit is set by defects caused by bad vacuum or grid emission. A valve in which the secondary emission current from the anode is large would not be suitable.

The power available in the anode circuit may be small, due either to the effective internal resistance being high or to the impulse being of very short duration. The output may be inadequate for operating certain types of counting circuit. The mixing circuit may in these circumstances be followed by an amplifying valve having sufficient output power or by an "impulse-lengthening" stage. An amplifying valve reverses the sign of an impulse; if this is undesirable a phase-reversing transformer may be used or the cathode follower circuit, which does not reverse the sign of the impulse, may be suitable.

Impulse lengtheners.

The three types of circuit commonly employed as impulse lengtheners are illustrated in Figs. 7·2, 7·3, 7·4. When a positive pulse is applied at the input of the circuit of Fig. 7·2 grid current flows so that when the pulse is over the grid charges up negative and stops the anode current for a time determined by the time constant CR_g of the grid circuit. A lengthened positive pulse therefore appears on the anode of the valve.

The action of the rectifier circuit of Fig. 7·3 should be obvious. The resistance R should be greater than the resistance of the rectifier when conducting. The capacity C is then chosen to make the time constant CR sufficiently long.

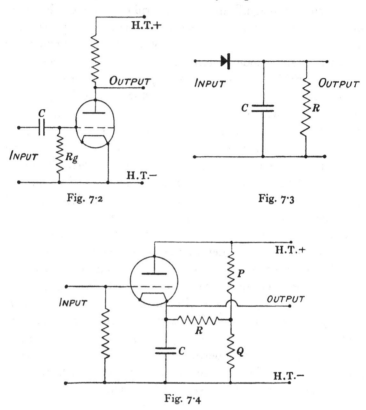

Fig. 7·2

Fig. 7·3

Fig. 7·4

In both these circuits a certain amount of power is required in the input circuit. The circuit of Fig. 7·4, however, requires a negligible power. The valve acts as a rectifier, the resistance of which is controlled by the grid. The length of the pulse is determined by the time constant CR', where $R' = R + \dfrac{PQ}{P+Q}$. The resistances P and Q form a potentiometer system to bias the valve, so that in the normal state it passes no anode current.

Triggered circuits.

The non-linear properties of valves give rise to the possibility of circuits having more than one stable state for the same applied voltages. The circuit may be changed from one stable state to the other by an impulse which brings it to the unstable state which must exist between the two stable states. Such an impulse therefore acts as a trigger. Such circuits find application as recording counters and as discriminators or peak voltmeters. They may be formed by combinations of two valves as in the multivibrator circuit (1) and in the flip-flop or zero frequency multivibrator (69), or with a single valve by employing the effects of secondary emission from one or more of the electrodes as in the dynatron. With regenerative coupling between the anode and grid circuits a single valve may have two stable states, one a steady state and the other in which an oscillation takes place for the same applied d.c. potentials. This phenomenon is described as oscillation hysteresis. The oscillation may also exhibit another property that it is automatically interrupted or blocked; this phenomenon is sometimes referred to as "squegging".

A gas discharge tube also possesses this property of two stable states for the same applied d.c. potentials. The striking voltage is higher than the burning voltage, so that at an intermediate voltage two stable states are possible, one with a discharge and the other with no discharge. A grid-controlled gas discharge tube or so-called gas-filled relay, gas triode, or Thyratron (the word Thyratron is a registered trade name (34)) has some advantages over the simple two-electrode tube.

The principles of these devices will first be discussed, and particular applications will be left for description later.

The gas-filled triode and thyratron.

If the grid is at the filament potential and the anode voltage is greater than the ionization potential, an arc is produced and a large anode current may flow. The pressure is low and an electron has a much longer free path than a positive ion. An electron has in fact a chance of being accelerated in the field to a velocity sufficient to ionize. If the anode voltage increased, this ionization, and consequently also the current, would be greatly increased.

When there is no arc, if a negative voltage is applied to the grid, it is necessary to apply a voltage much higher than the ionization potential to the anode to produce a field at the filament which will draw off electrons and thus strike an arc. Just as in a hard vacuum valve if $V_a + \mu V_g < 0$ no anode current flows, so with the gas-filled triode if $V_a + \mu V_g < 0$ no arc will strike. If now a positive impulse is applied to the grid, the arc at once strikes (it is necessary to limit the anode current to less than the filament emission by a series resistance in the anode circuit). Then it happens that if the grid returns negative it has no effect on the arc, in fact it has lost control. The reason for this is that a positive ion sheath is formed round the grid by the following mechanism.

When an electrode is put in a gas discharge and brought to a negative potential it is found that the positive-ion current flowing to it is independent of its potential; also it is covered by a non-luminous sheath whose thickness increases as the potential is increased negatively. This is explained as follows. When a current is carried by ions moving across an otherwise vacuous space the maximum current (i) which can flow is given by an expression of the form $i = Kv^{\frac{3}{2}}/d$, where K is a constant depending on the charge and mass of the ion and includes a numerical constant which may be determined from the geometry of the electrodes. It is assumed that the ions have no velocity other than that due to acceleration in the electric field. v is the potential difference between the anode and cathode and d is the distance between them. If then v is constant the maximum current passing depends only on d. Applying these considerations to the electrode at a negative potential placed in a gas discharge it is evident that the positive ions close to the electrode are attracted to it, but the maximum distance from which the ions can be drawn is given by $Kv^{\frac{3}{2}}/i$. In the equilibrium state the current i must be the rate at which positive ions enter the sheath which is a function of the discharge passing. Increasing v merely increases the distance from which positive ions are drawn, without greatly affecting the magnitude of the current. The discharge will not be appreciably affected beyond the limit of the sheath; if then the current is large v must be a very large negative potential to extinguish the discharge by drawing off the positive ions.

For this reason the arc cannot readily be extinguished by applying a negative potential to the grid; it is therefore necessary to interrupt the anode current for a time long enough to allow all the positive ions to diffuse to the walls and electrodes. This time may be shortened by putting a high negative potential on both anode and grid; it varies with these potentials, and according to the type of gas-filled triode may lie in the range from about 10^{-3} to 10^{-5} sec.

An important characteristic of a thyratron is its control ratio μ, which corresponds to the amplification factor of a hard vacuum triode. This control ratio is measured as the ratio of the increase of anode potential necessary to strike the arc to the increase (if small) of grid potential which also allows the arc to strike. If the grid is not uniform it is like several grids in parallel, that with the lowest μ will allow the arc to strike for the lowest potentials. This smallest μ must therefore be high, so that the anode must be very completely shielded by the grid. Thyratrons and gas-filled triodes are available with control ratios from 20 to 100. It is possible with certain gas-filled triodes to extinguish the arc by only a moderate negative potential on the grid.

It has been mentioned that the anode current of a thyratron must be kept considerably less than the emission from the cathode. The reason for this is that the cathode surface cannot withstand positive-ion bombardment and must be protected from this by a considerable electronic space charge. Moreover, it is important that the filament temperature should be maintained correctly as the emission varies rapidly with filament temperature. The life of a gas-filled triode may be only a few minutes if the filament temperature is too low.

The gas used is either mercury vapour or one of the inert gases, argon, helium and neon, or some combination of these. The mercury-vapour triode suffers from the disadvantage that the vapour pressure and hence the operating characteristics are sensitive to temperature.

The flip-flop circuit.

This is a symmetrical two-valve circuit which has two stable states (Fig. 7·5). The symmetrical state in which the anode currents of both valves are equal may be shown to be unstable, for if from this state the grid potential of one valve V_1 is increased its anode potential falls, reducing the grid potential of V_2 which increases the anode potential of V_2 thus increasing the grid potential of V_1 further. This process continues until the anode current of V_2 is reduced to

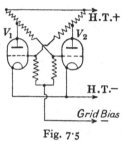

Fig. 7·5

zero. But the circuit is symmetrical, so it possesses two stable states in which the anode current of one or other valve is zero.

The multivibrator circuit.

The multivibrator circuit (Fig. 7·6) is somewhat similar to the flip-flop circuit, but the coupling from the anode of one valve

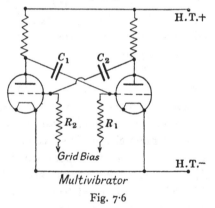

Multivibrator

Fig. 7·6

to the grid of the other is by a condenser only, so that the coupling is not maintained in a steady state. The circuit therefore oscillates. This oscillation may, however, be prevented by giving the two valves different grid potentials. It may be arranged that one valve normally passes no anode current, but if it receives a positive impulse on its grid, its anode passes a negative pulse to the grid of the other valve, and the rise of anode potential of this valve reinforces the initial positive pulse on the grid of the first valve. If the time constants $C_1 R_1$, $C_2 R_2$ are suitably chosen, this is maintained positive until the negative charge leaks away from the grid of the second valve. This is therefore a triggered impulse producer.

The dynatron.

This is essentially a negative-resistance device (33), but it is also non-linear and therefore may be used as a triggered device.

If an electrode of a valve is sufficiently positive, secondary electrons may be liberated from it by the primary electrons collected. If another electrode is maintained still more positive, it may collect the secondary electrons thus emitted. If the potential of the first electrode is increased, the secondary emission is increased. This increase of secondary emission may more than annul the increase of primary current. The nett current to this electrode therefore diminishes when the potential is increased. The slope resistance between the cathode and this electrode is therefore negative.

The transitron.

Another means of obtaining a negative slope resistance is illustrated in Fig. 7·7 (31). If the screen grid and suppressor grid

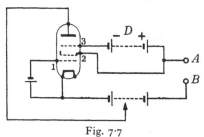

Fig. 7·7

of a pentode are connected together by a battery D, the current voltage characteristic between the points A and B is of the type shown in Fig. 7·8. The negative resistance portion arises because over this region the suppressor grid (grid 3) controls the division of the total current between the anode and screen grid in such a way that if it is made less negative the anode receives more current

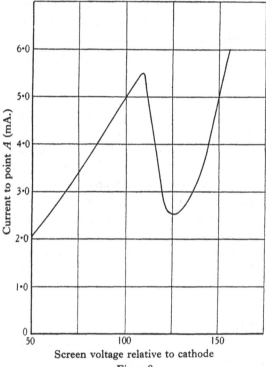

Screen voltage relative to cathode

Fig. 7·8

and the screen grid therefore less. It may be noted that over the negative resistance range the suppressor grid is negative with respect to the cathode and therefore draws no current. The battery *D* may be replaced by a condenser for a.c. operation; the bias on the suppressor grid is then derived from the cathode through a high resistance and a small battery if required.

Oscillation hysteresis.

If we have any ordinary valve oscillator and increase the negative grid bias, it is possible to apply a bias which if the valve were not oscillating would stop all anode current, yet the oscillation is maintained. If then the oscillation is stopped, by, for example, interrupting the anode circuit momentarily, it does not start again until the grid has been made more positive. This phenomenon is called "oscillation hysteresis".

Blocking oscillator or squegger.

It would, however, have been possible to stop the oscillation by applying a larger negative grid bias, and it is possible, by connecting a condenser paralleled by a high resistance in the grid circuit, to make the grid automatically charge up negatively by the grid current it draws during portion of the oscillation cycle so that the oscillation is stopped. The negative charge then leaks off the condenser and the oscillation starts again.

Alternatively, the normal grid bias may be too great for the oscillation to start again unless a positive impulse is applied momentarily to the grid. It is possible for the oscillation to be blocked during the first cycle of the oscillation.

This blocking oscillator is a very useful triggered impulse producer, particularly as the power in the impulse which trips it off may be negligible.

Discriminators.

Another non-linear application of valves is for discriminating between impulses of different voltages in such a way that impulses below a predetermined voltage are not passed on, whereas impulses exceeding this voltage by even a small amount are passed on as large impulses.

The simplest discriminator consists of a high-magnification valve with a large negative grid bias or cut-off bias, perhaps 100 V. (Fig. 7·9). A high resistance, 50,000 ohms to 1 megohm, is included in series with the grid to limit the grid current when large positive impulses are applied. The variation of anode voltage with the voltage applied at the input terminals is shown in Fig. 7·10 which makes the action of the circuit as a discriminator clear.

This circuit has four limitations:

(1) If the anode resistance is high, a very sharp impulse smaller than the cut-off bias may be passed to the anode circuit (via the grid-anode capacity of the valve) *without* change of sign. If the counter following the discriminator responds to such sharp impulses, this may be troublesome. A remedy would be to use a screen-grid valve, though the defect may be reduced by suitable choice of circuit constants where only slow impulses are to be counted.

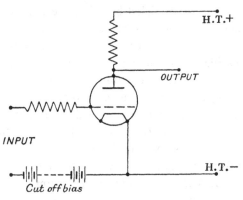

Fig. 7·9. Simple discriminator.

(2) It may be required to count slow impulses superimposed on a background having rapid variations such as is shown in the lower part of Fig. 7·10. If the counter has a very short resolving time the impulse shown would be counted as two, since it twice crosses the cut-off bias line from left to right. This defect cannot be completely eliminated with this type of discriminator, and it is probably most satisfactory to filter the input so that rapid fluctuations are suppressed.

(3) Due to the input capacity of the valve and the high resistance in series with the grid, very rapid impulses are diminished at the grid. For a similar reason the resolving time of the discriminator is limited.

(4) The flow of grid current also imposes a limitation. When impulses are arriving very rapidly the grid current flowing through

the grid-series resistance increases the effective cut-off bias. The discriminator is also not suitable for a resistance-capacity input coupling, since, owing to the charge collected on the coupling condenser, the effective cut-off bias is temporarily increased after each impulse.

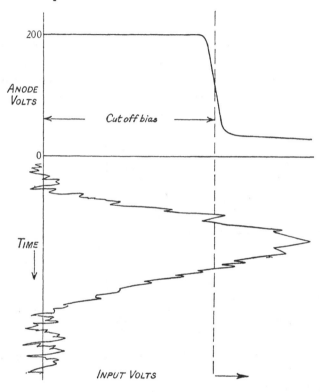

Fig. 7·10. Action of simple discriminator.

The impulses from such a discriminator are negative. For operating a thyratron counter it is convenient to follow the discriminator valve by an amplifier which reverses the sign and also produces impulses of a limited and constant size. There is no particular difficulty in this, but when this procedure is adopted the first-mentioned defect of the discriminator is aggravated and it will be almost essential to use a screen-grid valve as discriminator. Preferably this should be a pentode of a type in which the screen-grid voltage may be the full high-tension voltage, since it is unsatisfactory to supply the screen-grid voltage from a high-resistance potentiometer, for in that case the mean potential would depend on the rate of counting.

Many other discriminators have been used which obviate some of these disadvantages, but it is only recently that a discriminator has been developed which is almost perfect in performance. It will operate with impulses as slow as desired or as short as a microsecond without any readjustment. It will discriminate in-

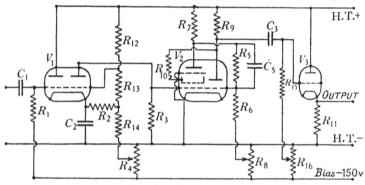

Fig. 7·11. Discriminator.

Values: R_1, R_{10}, 500,000 ω; R_2, 5000 ω; R_3, 50,000 ω; R_4, 10,000 ω; R_7, R_5, 250,000 ω; R_6, R_8, R_9, R_{15}, R_{16}, 100,000 ω; R_{11}, R_{13}, R_{14}, 20,000 ω; R_{12}, 3 MΩ; C_1, 0·01 μF.; C_2, optional; C_3, C_5, 0·0001 μF.
Valves: V_1, 6N7; V_2, AC/TP Mazda; V_3, MH41 Osram.

fallibly between pulses differing by only $\frac{1}{2}$ V. Impulses may exceed the cut-off bias by about half the high-tension voltage (200 V.) without drawing grid current or in any way disturbing the action. The output pulse may be controlled by the circuit constants to have a minimum duration as desired from 10^{-5} to 10^{-2} sec. or possibly a greater range.

The circuit of this discriminator is shown in Fig. 7·11. The first triode section of the first valve is effectively a cathode follower with the cathode normally maintained at a higher voltage than the grid by the cut-off bias. The cathode potential is constant until a voltage greater than the cut-off bias is applied across R_1.

Any further increase of this voltage raises the cathode potential. This renders the second triode section of the valve non-conducting, so the anode voltage rises to that of the negative high-tension line, thus removing the bias from the control grid of the pentode section of the next valve. This second valve is connected in a type of flip-flop circuit. Normally the triode section is conducting, so that its anode potential and consequently the screen-grid potential of the pentode section is low. A negligible anode current is flowing in this pentode section due to this low screen-grid potential and the bias on the control grid. The anode potential is therefore high, and this through R_5 and R_6 maintains the grid bias on the triode at a low value. Under the action of a positive impulse on the control grid of the pentode, the anode potential falls, increasing the negative grid bias on the triode section and thus raising the anode potential and the screen-grid potential of the pentode section which further increases the current through the pentode section. At a certain bias on the control grid of the pentode section this change takes place abruptly, providing a sharp output impulse even if the input impulse is extremely slow. The reverse change does not take place until the input voltage has fallen considerably below that which first triggered the circuit, as the screen-grid voltage is now much higher. Consequently, the circuit is immune from being triggered by small fluctuations superimposed on an impulse.

If the negative impulse which occurs on the anode of the triode when the reverse change takes place is undesirable, it may be suppressed in the output or by a cathode follower valve biased to be normally non-conducting.

Since R_5 may be a high resistance, it is bridged by the condenser C_5 in order to retain the sensitivity for impulses of very short duration.

Another facility provided by this discriminator is found very convenient. It is very desirable that a high-speed counter should also count accurately at very low speeds. Now it so happens that conditions are often not very favourable, and when counting at a very low speed a slight disturbance, such as microphonic trouble in the ionization chamber, causes the counter to record a rapid burst of impulses which necessitates cancelling the observation and possibly leads to a considerable waste of time. It is therefore desirable to have some control by which the resolving time may be lengthened very considerably when counting at low speeds, so that chance disturbances can only add to the count a small number which may even be determined by maintaining a monitoring watch by ear. This control is provided by adjustment of C_2 or, if a very long resolving time is required, also by increasing R_2, R_3 and R_{14}.

The original impulse on the grid C_1 charges up the condenser C_2 through the valve; this charge can only leak away through R_2 and R_{14}, so by making the time constant $C_2 R_2$ large any impulse on the cathode and hence on the anode A_2 must be long. A microphonic disturbance is liable to cause the first triode section to conduct repeatedly, but this only superimposes a fluctuation on the output impulse which does not cause the triggered circuit following to record more than one impulse. It should be noted that if R_2 and R_{14} are made large R_3 must also be made large to limit the anode current of the second triode section which also flows through R_2 and R_{14}.

The resistances R_{12}, R_{13} form simply a potentiometer system for maintaining the normal bias potential of G_2. The resistance R_{14} provides negative feedback to limit the amplification of this triode section to about two. This preserves the discrimination of the second valve against superposed background on an impulse.

In conclusion, it may be remarked that a thyratron counter has often been used as its own discriminator. Many of the cheaper gas-filled triodes now available are not very suitable for this use, as they may be extinguished by a large negative grid potential, so the discharge will only pass for a portion of the duration of the impulse. In practice grid bias up to 20 or 30 V. may usually be applied. The striking potential may be regarded as constant within about 3 V., so that if the impulses to be counted are well above the background and not very different in size, a separate discriminator is unnecessary.

Chapter VIII

RECORDING COUNTERS

The problem of the mechanical recording counter is not essentially difficult; it has been governed in the past—and still is—by what is commercially available rather than by what is physically possible. What is generally required is an electromechanical device which counts up to 10,000, and which operates in the shortest possible time on the power which is readily available in the anode circuit of an ordinary valve. Probably no such device which even approaches the mechanical limit has yet been constructed. Electromechanical devices responding up to frequencies of the order of 20,000 c./sec. without depending on resonance effects are to be found constructed on the lines of certain loud-speakers; the amplitude of movement is, however, rather microscopic at these frequencies. Electromechanical relays have been made in which the operation is completed in $\frac{1}{2000}$ sec. or less. These figures may give some indication of where the limit may lie, but it may be significant also to note that what is required of a recording counter is merely a visual indication. A beam of light need add nothing to the inertia of the device, and, moreover, microscopic movements are permissible. It is probable that ease of observation, cheapness and reliability will condition the practical limit.

Telephone-message registers and selector switches.

Probably the most commonly used device is the telephone-message register, such as is used in telephone exchanges for counting the number of calls put through by a subscriber. Quite a number of patterns are available (68), and on the average they operate in $\frac{1}{10}-\frac{1}{30}$ sec. and require from 0·7 to 5 W. for operation; the greater the power expended the shorter the time of operation can be made. Resistance values from 300 to 4000 ohms are available.

A few years ago there appeared on the secondhand market step-by-step selector switches electromagnetically operated, such as are used in totalizators and automatic telephone exchanges. Available thus cheaply they have been applied as electro-mechanical recorders, for the time of operation may be reduced to $\frac{1}{120}$ sec. In order to obtain this it is necessary to operate the recorder in series with a relatively high resistance, as an electrical limit to the speed of operation is encountered, namely, the time constant L/R of the winding. The power required for the high speed of operation is about 0·6 amp. at 200 V. These selector switches have been designed for quick operation but are heavily loaded by the switch arms; these should therefore be removed for economy of power.

The limit to the speed of operation of telephone-message registers is largely set by the moment of inertia of the armature and the maximum force which can be developed on the armature. This is limited by the magnetic leakage, the saturation of the magnet core and the heating of the winding. The operating time may be reduced to its minimum by stiffening the return spring and limiting the travel of the armature as much as possible. Many counters have a catch with a gravity control; for high-speed operation this should be replaced or augmented by a spring control.

In many circuits the counter is required to release itself after operation. This may be done by the armature breaking a contact in series with the winding or making a contact which short-circuits the winding. The momentum of the armature and the inertia associated with the magnetic field usually makes a simple device of this kind mechanically satisfactory. Electrically this contact is liable to be troublesome, for it is in a highly inductive circuit, of which the inductance varies with the position of the armature. Sparks at make and break cannot therefore be completely suppressed and the counter is liable to give electro-magnetic pick-up in other parts of the circuit. A condenser of about 1 μF. in series with a resistance of 100 ohms connected between the contacts is generally found a fairly satisfactory suppressor. Alternatively, this may be connected across the winding of the counter and a smaller condenser, $0.1-0.02\ \mu F$., may be connected across the contacts.

Sparks may also be suppressed by disks of silicon carbide specially fired and sold under various trade names such as Thyrite, Metrosil, Atmite. These disks have a very high resistance for a low applied voltage, but as the voltage is raised the current rises and this causes a very marked instantaneous decrease in resistance so that the current increases very considerably. The disks are usually connected in parallel across inductive windings.

Fast-counting meters.

There is one meter on the market which may be included in the category of fast-counting meters. It is the Cenco (15) meter which ordinarily operates in $\frac{1}{120}$ sec. but which may be made to operate in $\frac{1}{200}$ sec. if a high-voltage impulse is applied (66). It is not easy to read, since it records only up to 60 for one revolution of the main pointer, and a small subsiduary pointer indicates the total number of revolutions. A scale-of-ten counter operating a telephone-message register costs little more to construct than the Cenco meter together with its operating circuit. The latter, though much inferior in performance, is, however, simpler.

A number of fast-counting meters have been made using the mechanism of a watch with the escapement replaced by an electrically operated release. Accounts of some of these have been published (23, 43, 75). A minor objection to these is that they require periodical winding up.

The fastest counting meter yet described is that of Neher (50). This has been made to operate in $\frac{1}{2000}$ sec. This achieves its speed by the use of a very small ratchet and very light moving parts. Fast-counting meters have also been made at the Cavendish Laboratory which operate reliably in $\frac{1}{400}$ sec. and operate two hands which allow direct readings up to 1000 without any complicated conversions of scales. These meters also depend for their speed on a very small ratchet and low moment of inertia of the moving parts. A sketch drawing of one of these is shown in

Fig. 8·1. The armature A is attracted by the electromagnet M. The paul P advances the ratchet while Q slides over one tooth. On the return stroke the paul Q continues the motion of the ratchet while P slides over a tooth. The movement of the ratchet

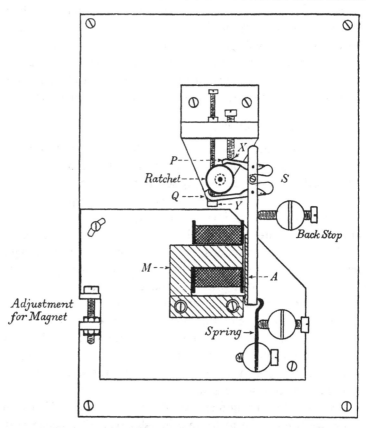

Fig. 8·1. Full size.

is thus spread over the whole time of operation so as to minimize the accelerations and decelerations. Stops X and Y are arranged so that at no point of the stroke can the ratchet step on by an extra tooth. The pauls slide past these stops until they drop over a tooth; one of the pauls is therefore always sliding between one tooth of the ratchet and a stop, so the ratchet cannot advance by an extra tooth. The ratchet has fifty teeth. The pauls and the teeth are of hardened steel, and are designed so that considerable wear can be permitted. The limit of all such high-speed counters is the compromise between permissible wear and lightness of the moving parts. The hands are driven through a very light flexible spring coupling so that the inertia of the hands and gearing is not added to that of the ratchet. The main hand is of $1\frac{1}{4}$ in. radius and is geared by reduction gearing of ratio 20 : 1 to a concentric hand of $\frac{3}{4}$ in. radius. This hand therefore makes one revolution for 1000 impulses. The resistance of the winding is 800 ohms having 5000 turns of 44 s.w.g. wire. The voltage impulsively applied is about 200.

High-speed counters.

It will be explained later that it is often desirable to have a counter which is able to resolve, that is, to count separately, impulses which are only 10^{-4} sec. apart or even less. In view of this, electrical recording circuits capable of such resolution have been devised. These operate in such a way that only every tenth or every eighth impulse is passed on to an electromechanical recorder. The general principle most commonly employed is that of the scale-of-two counter introduced by Wynn-Williams. Each successive stage in such a counter divides the counting rate by two. The second stage is operated by every second impulse, the third stage by every fourth impulse; the electromechanical recorder or a fourth stage is then operated by every eighth impulse.

In the original Wynn-Williams counter (80) the unit which passes on one impulse for every two it receives is a simple symmetrical circuit employing two thyratrons (Fig. 8·2). An arc is struck in one thyratron the grid of which therefore loses control; a positive impulse applied at the input will, however, cause an arc to strike in the other thyratron. Its anode voltage therefore drops from the supply voltage (200 V.) to the arc voltage (15 V.). This drop of potential is communicated to the anode of the first thyratron which therefore drops from 15 V. to nearly -170 V., extinguishes the arc, and the anode potential then rises to the supply voltage. The effect of the impulse has thus been to transfer an arc from one thyratron to the other. The large rise of voltage on the anode of one thyratron after the arc has been extinguished is used as an impulse to supply the next unit.

It should be clearly under-stood that no circuit yet proposed for a scale-of-two counter is infallible, and, al-though in ordinary use possible failings may be unimportant and the counter may appear reliable, a proper appreciation of the peculiar limitations of the counter in use is necessary in any work in which the exact operation of the counter is relied upon. The following

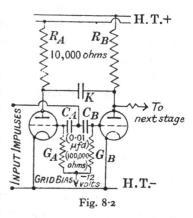

Fig. 8·2

discussion of the limitations of different counting circuits is not therefore to be taken as indicating that the counter is necessarily troublesome or unreliable.

The original scale-of-two unit has the merit of simplicity and, provided it is not worked near the permissible maximum speed or with impulses of irregular form, may be considered quite reliable. With a high rate of counting, however, a second impulse may arrive before the anode potential of the thyratron just extinguished has risen to the full value; the thyratron may, nevertheless, strike. A similar effect may occur with a single impulse if it is protracted and irregular. This may upset the normal action of the counter in two ways. In the first place it is the rise of anode potential of one of the thyratrons which operates the next unit. If then this rise is curtailed, the impulse passed on may be insufficient, and in this way a close pair of impulses may appear to have been missed by the counter. In the second place, when one of the thyratrons strikes before its anode potential has risen to the full value, the fall of anode potential communi-cated to the other thyratron is less than the normal and may be insufficient to extinguish the arc. Both thyratrons remain alight and the counter is completely jammed.

These two defects were obviated (37), the first by providing a stage of valve amplification between the first and second units so that even a small impulse does not fail to operate the second unit; the second by the addition of condensers C_{AG} as shown in Fig. 8·3. The fall of anode potential when one thyratron strikes is communicated by this small condenser to the grid of the other. The grid potential is thus made strongly negative, and the thyratron

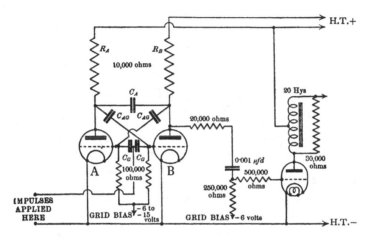

Fig. 8·3

is prevented from striking until the charge has leaked off its grid condenser. In the meantime the anode potential of the same thyratron rises to the full value. It is found with this circuit that it is possible to use a much smaller extinguishing condenser C_A, so that the anode potential rises more rapidly after extinction and the resolving time is reduced. This is due to the fact that the negative potential on the grid hastens the collection of positive ions and thereby shortens the deionization time. The stage of valve amplification between the units is essential with this circuit to prevent impulses being fed back from the next unit.

Circuits using hard vacuum valves.

It is evidently possible to make the flip-flop circuit act as a scale-of-two counter, as it has two stable states. It is only necessary to provide some means of triggering the circuit alternately from one stable state to the other. A number of circuits have now been published in which this has been achieved. These must be judged by their simplicity, performance, and reliability.

The circuit for which the greatest claims of performance and

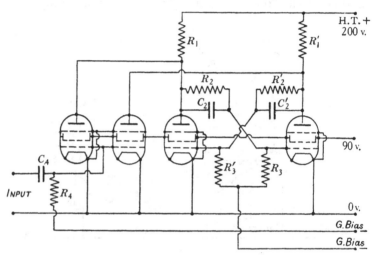

Fig. 8·4. Stevenson and Getting's scale-of-two unit. All valves Type 57.
R_1, R_1', R_3, R_3', R_4, 100,000 ω; R_2, R_2', 300,000 ω; C_2, C_2', 0·00025 μF.;
C_4, 0·000025 μF. Resolving time less than 2×10^{-5} sec.

reliability have been made is that of Stevenson and Getting (72) (Fig. 8·4), but unfortunately this circuit uses four pentodes per stage. The high cost of pentodes in this country hinders its adoption. Apparently this circuit is capable of resolving impulses only 2×10^{-5} sec. apart, and retains this property over the whole of the very large range of grid bias which is permissible.

The simplest of the circuits is that published by Alfvén (3) (Fig. 8·5, p. 85), and also independently by Lifschutz and Lawson (41). This circuit is perfectly satisfactory provided it receives sharp impulses of controlled magnitude. It has not, however, proved so satisfactory when the resolving time is reduced to the minimum.

It has proved advisable to check the action of any counter circuit in detail, before trusting its operation near its limit. That is to say, tests should be made of the resolving time, the certainty of passing on an impulse to the next stage or thyratron, when a close pair of impulses is received, the range of impulse forms which it can handle, and how these properties are retained over the grid-bias range.

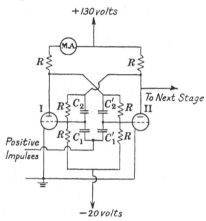

Fig. 8·5. Alfvén's scale-of-two unit.

Valves I and II, Philips B2038 ($\mu = 33$, $S = 3\cdot35$ mA./V.). All resistances $R = 0\cdot2\,M\Omega$. C_1, C_1', 1000 cm.; C_2, C_2', 500 cm. For high-speed counting smaller condensers are preferable.

The circuits published by the writer (38) (Fig. 8·6, p. 86) are complicated by the inclusion of cuprous oxide rectifiers (Westectors) and double-wound chokes. The particular objection to these chokes is that they are components which cannot be tested and measured so readily as condensers and resistances by anyone with restricted laboratory facilities. The writer has set up other circuits avoiding the use of such components but has not found any such circuit without some disadvantages. The ideal simple circuit which can be set up by anyone from a simple written specification using only two valves per stage of any make, the other components being only resistances and condensers which can be relied upon to work without fail, has yet to be devised (56).

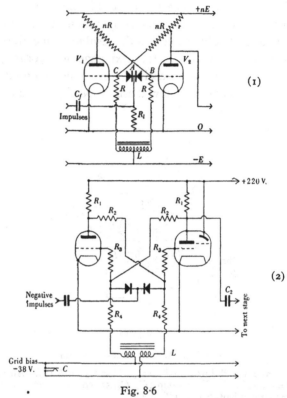

Fig. 8·6

Circuit (1). $V_1 = V_2 =$ Mazda L2; $AB = AC =$ Westector, Type W6; $r = 40,000 \omega$; $nR = 240,000 \omega$; $R = 100,000 \omega$; $R_l = 100,000 \omega$; $L = 60$ Hy; Centre tapped; $C_f = 0.001 \mu$F. Counts impulses of duration $> 2.5 \times 10^{-5}$ sec. < 0.01 sec. Resolving time 2×10^{-4} sec.

Circuit (2). Valves Mullard 904V and TV4 (tuning indicator). $R_1 = R_2 = 250,000 \omega$; $R_3 = R_4 = 100,000 \omega$; Rectifiers, Westectors Type WX6. $C_1 = C_2 = 0.0002 \mu$F. $L =$ Two $2,000 \omega$ headphone bobbins on centre limb of two E-shaped stampings of Laminic. Counts impulses of duration up to 0.01 sec. Resolving time 10^{-4} sec.

These remarks show that the nature of the difficulty is not such that it would trouble a manufacturer or anyone making a number of counters. Research apparatus of this type must, however, be virtually home constructed.

Indicators.

It is necessary in these circuits to provide some means of indicating in which stable state the circuit is at any time. The most attractive means of doing this is to use a cathode-ray tuning indicator in place of one of the valves of each pair. This, however, imposes a restriction on the circuit, for the triode portion of these indicators is usually limited to working with very small currents and consequently high resistances in the circuit. This makes the circuit unsuitable for a resolving time less than 2×10^{-5} sec. Apart from this limit, which in most cases is quite unimportant, very satisfactory counters have been made using such indicators.

Alternatively, small neon-lamp indicators may be used. It must be remembered, however, that the ignition voltage of a neon tube is likely to increase with age, and an ample reserve voltage must be provided in the circuit. Also, if the applied voltage is somewhat low the discharge may not strike immediately the voltage is applied. The striking of the tube then produces another impulse in the circuit and this must not be passed through to operate the counter in any way.

A milliammeter in one of the anode circuits does not prove quite so convenient an indicator, since the time lag in its operation renders it useless for visual checking of the action of the counter at high speed.

Action of the circuits.

The action of the various circuits may be briefly described.

In Stevenson and Getting's circuit (Fig. 8·4) positive impulses are applied to the control grids of two pentodes connected together. The common grid bias is normally such that no anode current flows. An impulse causes both pentodes to pass anode current. The two anodes are separate and connected directly to the anodes of the second pair of pentodes which are connected in a simple flip-flop circuit with the addition of condensers bridging the anode-to-grid resistances as in the multivibrator. The use of pentodes is necessary to maintain the very short resolving time. The screen grids are all maintained at a steady potential (90 V.).

One of the pentodes in the flip-flop circuit will be passing anode current so that its anode potential will be low; the extra current drawn by the impulsing pentode will not lower this anode potential greatly. The other pentode in the flip-flop circuit will be passing no anode current, so its anode potential will be high until the impulsing pentode draws anode current. The potential of the two anodes will then fall and a negative impulse is communicated by a condenser C_2 to the grid of the other pentode, thus causing the system to change to its other stable state. The system is symmetrical, so that, after the steady state has been reached, another impulse would again reverse the conditions. The action would be upset by a prolonged impulse which would maintain all the anodes simultaneously at a low potential. It is arranged that only short impulses can be transmitted by making the time constant of the input circuit $C_4 R_4 < \frac{1}{5} C_2 R_2$. The latter time constant $C_2 R_2$ approximately represents the time constant for the system reaching its stable state after a transition.

In Alfvén's circuit (Fig. 8·5) positive impulses are applied. Suppose that the grid of valve 1 is positive so that the valve is conductive, while the grid of valve 2 is negative so that valve 2 is passing no anode current. The resistance of the grid-filament path of valve 1 is small and the applied positive impulse has little effect, therefore, on the grid voltage and hence on the anode voltage. The grid input resistance of valve 2, on the other hand, is very high, so that the positive impulse produces a large change in grid voltage and produces a negative impulse on its anode. This is transferred to the grid of valve 1 by the condenser C_2 and, being larger than the applied positive impulse, puts the grid of valve 1 negative. The resulting positive impulse on the anode of valve 1 increases the positive impulse on the grid of valve 2 through the condenser C_2'. This change-over takes place just as in the multivibrator circuit, the resistances cross-connecting the anodes and grids act as in the flip-flop circuit to maintain the stable state which has thus been reached.

The circuit can also be triggered by large negative impulses; such impulses must therefore not be allowed to occur in the input. This may be done by a rectifier as in the output of the last discriminator described in Chapter VII.

Two of the counting circuits described previously by the writer(38) are shown in Fig. 8·6. In circuit (1) of Fig. 8·6 negative impulses are applied at A. If, in the initial state, valve V_2 is passing no anode current, A is almost at the same potential as C and is positive with respect to B. The rectifier AB will therefore be in a non-conducting state, and very little of a negative impulse at A is passed on to B, whereas a large fraction of the impulse will be passed on to C through the rectifier AC across which there was initially a negligible difference of potential. The negative impulse thus reaching the grid C causes the anode current of V_1 to fall; the potential of B therefore rises, and if the impulse is sufficiently large B will reach the potential of C. In this state the currents in the two symmetrical halves of the circuit are equal. This condition is unstable, and from it the system may pass to either of its stable states, but the inductance L is included to continue the change of relative potential of B and C so that the stable state which results is not that from which the system started.

The circuit (2) is slightly different in that the centre point of the two rectifiers is left to find its own potential, which will be approximately that of the more positive grid, so the mode of operation of the circuit is still as has been described above.

The performance of the circuits is indicated briefly with the specifications below the figures.

Three scale-of-two units followed by an electromechanical counting meter has been found most generally useful. The counting meter operates for every eighth impulse and, if an ordinary telephone-message register is used, its change of reading must be multiplied by 8 and the odd number up to 7 added in from observation of the state of the counter. This may be read quite easily from the indicators in the three stages if they have been set to zero before starting the count, for the first may be labelled 1, the second 2 and the third 4, the corresponding numbers being added in to the count if indicated. To avoid the multiplication by 8 the meter may be modified to read directly in multiples of 8. This may be done by altering the figures on the units drum from 0123456789 to 0864208642, and by fitting additional teeth so that the tens drum is moved on at every movement of the units drum except when this moves from showing 0 to showing 8. The meter thus shows the sequence 008, 016, 024, 032, 040, 048, ...

Scale-of-ten counters.

It would be still more convenient if the electrical counter operated on a scale of ten. Wynn-Williams devised a thyratron ring counter which could have any number of thyratrons arranged in a ring with an arc in one of them, but this proved difficult to adjust. It has proved possible to extend the basic idea of the flip-flop circuit to five valves, only one of which passes no anode current. These may be linked in a ring by condensers, and condensers and resistances, so that impulses cause the system to pass to each of its five stable states in rotation. By using cathode-ray tuning indicators instead of ordinary valves it is evident which valve is passing no anode current. By placing this ring-of-five to follow a scale-of-two unit, a scale-of-ten counting meter may be operated directly to record tens, hundreds, and thousands, the units being read directly from the ring-of-five and the scale-of-two unit.

The circuit is shown in Fig. 8·7. Positive impulses are applied simultaneously to all the grids via the separate small condensers. The grid-filament resistance of those valves whose grids are positive is, however, low, so that the impulse on the grid is small; the negative impulse received from the anode of the previous valve in the ring is also small except for that following the valve which was passing no anode current. The positive impulse on the grid of this latter valve is large and its anode potential falls. The negative impulse passed to the next valve is sufficient to stop its

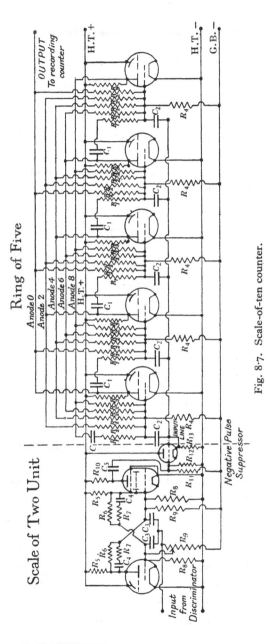

Fig. 8·7. Scale-of-ten counter.

Values: R_1, 2 MΩ; R_2, 1 MΩ; R_3, R_5, R_6, R_7, 250,000 ω; R_4, R_9, 500,000 ω; R_8, R_{13}, 100,000 ω; R_{10}, 140,000 ω; R_{11}, 2 MΩ; R_{12}, 40,000 ω; C_1, C_4, 0·001 μF.; C_3, C_2, 0·0005 μF., C_5, 0·0003 μF.

anode current. The flip-flop connexions make this new state stable and it persists until another impulse is received.

Circuits for operating mechanical recorders.

The simplest circuit for operating a telephone-message register is probably a single thyratron circuit (Fig. 8·8). A break contact is fitted to the message register to interrupt the thyratron anode circuit. The means adopted for suppressing sparking at this contact have already been discussed. It may, however, be added that the insulation of the condenser must be high or sufficient leakage current may pass to maintain the arc in the thyratron.

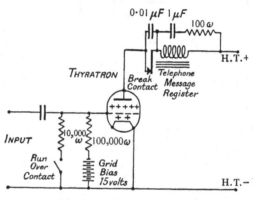

Fig. 8·8. Thyratron counter.

With message registers requiring a small enough current, a type of flip-flop valve output circuit may be used. One such circuit is illustrated in Fig. 8·9. Again, the counting meter is fitted with a break contact B, but in this circuit it interrupts only a very small current, so the contact is not troublesome to keep clean nor is there any need for special measures for spark suppression.

Very satisfactory valve circuits may, however, be made which avoid the need for any contact by providing only a pulse of current. The magnitude of this pulse may be adjusted to be sufficient to operate the counter without fail. A circuit which has proved very generally satisfactory for operating all types of

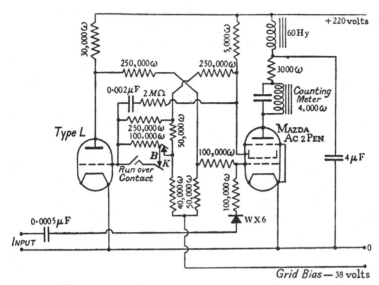

Fig. 8·9. Flip-flop meter operating circuit. Contact K is ópened in the home position of the counting meter.

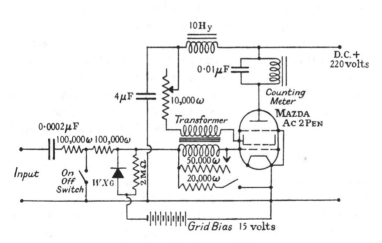

Fig. 8·10. Blocking oscillator.

counter, including fast-counting meters, is shown in Fig. 8·10. This is a blocking oscillator circuit. The triode formed by the cathode, control grid and screen grid is connected as an ordinary oscillator with regenerative coupling provided by the transformer between the screen-grid and control-grid circuits. A large negative grid bias is, however, applied to the control grid which stops all anode current. The circuit is triggered by a short positive impulse which momentarily removes the negative bias. The circuit would then maintain an oscillation, but the damping resistance across the secondary of the transformer in the control-grid circuit ensures that after one cycle of the oscillation the control grid never returns sufficiently positive to permit another cycle to occur. The pulse of current in the anode circuit is used to operate the counter. The duration of this pulse is controlled by adjusting the time constant L/R of the screen-grid circuit, and by adjusting the damping resistance. The main control of the pulse duration is, however, effected by suitable choice of the transformer. For example, for pulses from $\frac{1}{40}$ to $\frac{1}{10}$ sec., suitable for operating a telephone-message register, the transformer may be a good-quality audio-frequency inter-valve transformer such as the Ferranti A.F. 3 (primary inductance 220 H. (60 H. at 6 mA. d.c.), step up ratio 1 : 3·5). For pulses of $\frac{1}{100}$ sec. or less an audio-frequency output transformer such as the Ferranti O.P.M. 5 (primary inductance 70 H. (6 H. at 50 mA. d.c.), step down ratio 4 : 1) is suitable. The use of a Westector rectifier in the grid circuit maintains a high input resistance while minimizing the dead space or time interval after an impulse during which the counter remains insensitive. This dead space should not, however, be too short as the counter should be allowed sufficient time to release. This may be tested by setting the grid bias to zero: the counter should then operate repeatedly or "run over" at nearly its maximum speed.

It will be noticed that in the circuits shown for operating counting meters a "run-over" contact is shown; this may be used for running the counter on to the next hundred or thousand in order to start each count from a round number. A contact operated by the counter may be arranged to interrupt the run-over contact circuit in this home position, that is, when the hundred or thousand is reached. If the meter runs over somewhat slowly, relay circuits may be arranged for this operation, but with the fast-counting meters the run-over contact is simply a push button which is pressed until the meter reaches its home position, when it stops running over, the push button then being released.

Counting-rate meters.

For some purposes it is advantageous to have a direct indication of the rate of arrival of particles. Owing to statistical variations in the number arriving in any given time interval such an indicator cannot be very quick acting unless the rate is very high. The principle may, however, be quite simple. Impulses of strictly the same amount are derived from the particles and are fed into a condenser across which there is a constant leak resistance; the average potential difference maintained across the condenser is then a measure of the rate of arrival of the impulses.

One of the remarkable features of the technique of counting is that it proves possible to derive results from rates as low as one particle in a few minutes, and with the same apparatus to count thousands of particles per minute. This range of tens of thousands to one has proved invaluable in certain investigations, but for many experiments a much smaller range such as could be read on an ordinary direct-reading instrument is sufficient. A counting-rate meter which can measure rates from 30 to 6000 particles per minute finds much application.

Such a meter has been devised and applied by Gingrich, Evans and Edgerton [27]. Their apparatus consists essentially of two parts: a uniform impulse producer and a resistance-capacity tank circuit for averaging the number of impulses over a time of the order of half a minute.

This capacity-resistance averaging circuit is connected in the anode circuit of a pentode so that the magnitude of each pulse supplied to the capacity is independent of the voltage across it. The uniform impulses are supplied to the grid circuit of the pentode. A microammeter in series with the resistance measures the voltage across the condenser. Gingrich, Evans and Edgerton use a condenser $C = 100\,\mu\text{F}$. and $R = 3 \times 10^5$ ohms and the meter reads $0\text{–}200\,\mu\text{A}$. Alternatively, they suggest using a smaller condenser with a valve voltmeter to measure the voltage. A convenient valve voltmeter circuit which the writer has used for this purpose has been described in Chapter VI.

Two impulse generators were proposed by Gingrich, Evans and Edgerton, one essentially the first stage of a thyratron scale-of-two counter, the other an unbalanced multivibrator circuit. No information is given of the uniformity of the pulses produced nor of the dependence of this on the form and magnitude of the input impulse nor how this uniformity depends on the resolving time.

O. Viktorin and the writer (unpublished) have applied a blocking oscillator circuit similar to that of Fig. 8·10 for producing uniform impulses which are supplied to a condenser ($8\,\mu\text{F}$.) shunted by a high resistance (2·5 megohms) connected in the anode circuit. The transformer used is very small, so that the duration of a pulse is only about 3×10^{-4} sec. The voltage developed across the condenser is read with a valve voltmeter using the circuit of Fig. 6·3.

It should be pointed out that the counting-rate meter is of less general application than a counting circuit. It possesses a decided advantage for taking a continuous record of a slowly changing rate, but for other purposes where the rate is frequently changed the counting-rate meter necessarily requires a longer time to obtain a result of a given statistical value.

Count integrators.

A recording counter which records with certainty any number from one to several million if required cannot seriously be challenged on performance. For some purposes, however, it would be an advantage to make simultaneously a large number of counts, say twenty or more. An installation of twenty high-speed counters would be somewhat elaborate and costly. In such circumstances a device which may be described as a count integrator may prove useful. The arrangement is similar to that of a counting-rate meter except that the leak resistance across the storage condenser is omitted. Twenty separate circuits feeding twenty storage condensers may readily be set up; the charge accumulated on each condenser during the counting time may subsequently be discovered by testing the condensers in succession with a single-valve voltmeter.

A similar principle has been applied to make a scale-of-ten counting circuit. The condenser receives and stores nine impulsive charges, then on receiving a tenth charge it triggers a circuit which discharges the condenser. This might seem a simpler system than the scale-of-two and scale-of-ten counters previously described, but the electronic means of discharging the condenser introduces complications which make the two systems comparable.

Chapter IX

GEIGER-MÜLLER TUBE COUNTERS

The Geiger-Müller tube counter operates by the production of an electrical discharge in a gas. The counter is extremely sensitive, and a discharge may be produced when a single pair of ions is liberated almost anywhere within the tube. It is distinguished from other discharge counters by this large volume over which the formation of an ion pair will produce a discharge.

The counter in its usual form consists of a cylindrical metal tube, along the axis of which a thin wire is stretched. The wire is usually bare or slightly oxidized and is highly insulated from the metal tube. The tube is customarily about 10 cm. long but may be five or ten times longer or shorter. Most satisfactory operation is secured when the length of the tube is greater than about twice its diameter. The tube is filled with gas at a low pressure, most commonly between 5 mm. and 20 cm. of mercury. The wire is maintained at a positive potential with respect to the tube.

These counters are used in large numbers, particularly in cosmic-ray work. It is still, however, impossible to guarantee that a counter made up carefully to any specification will be satisfactory in all respects. The extreme sensitivity of the counter renders it liable to apparently spontaneous discharges, the origin of which is uncertain. The characteristics of the discharge are also often found to change with time.

If no restriction is placed on the shape, size and material of the counter, the nature and pressure of the gas, the operating voltage and the diameter of the wire, it seems possible to produce counters of which the great majority will operate at least qualitatively in a satisfactory manner.

The varied nature of the discharges in Geiger-Müller counters made to different specifications and operated under different conditions should be clearly recognized. Generalizations about the behaviour of counters are of little value unless the operating conditions and the type of the discharge are specified. In discussing the operation of counters it is necessary to make a clear distinction between three processes occurring in the counter: the initiation of the discharge, its growth, and its extinction.

Initiation of the discharge.

Suppose for the present the growth and subsequent extinction of the discharge are taken for granted. Evidence about the initiation of the discharge is mainly obtained from characteristic curves of the type shown in Fig. 9·1, which shows the number of

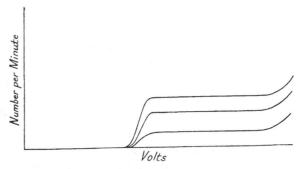

Fig. 9·1. Geiger counter "ideal" characteristic.

discharges per minute plotted against the voltage applied to the counter under the action of a constant weak source of ionization.

The various curves of Fig. 9·1 refer to ionization sources of different strength. They have the form which is claimed to be the ideal characteristic for a Geiger-Müller counter. As the voltage is increased the number of counts is first zero, then rises steeply to a flat portion where the rate of counting is independent of the voltage. At a higher voltage the number of counts again increases, this increase in the ideal case being independent of the source of ionization.

On these curves only two causes for the initiation of the discharge are indicated: initiation by ionization, and spontaneous initiation which is always obtained at high voltages and sometimes also at the working voltages. Included in the initiation by ionization is the "natural" of the counter, namely, the discharges produced by radioactive contamination of the counter and by cosmic rays. The spontaneous discharges are sometimes vaguely attributed to the presence of irregularities in the surface of the wire. Sharp points on the wire certainly lower the discharge voltage and therefore produce spontaneous discharges at a lower voltage. There is, however, no evidence that this is even a frequent cause of the defect. The writer has had satisfactory counters in which the wire, of iron, was covered with rust crystals. It is possible that spontaneous discharges are initiated by the local concentrations of electric field which occur at the junctions between the conductors and solid insulating materials, as the field at such points is liable to fluctuations.

An unfortunately common defect of Geiger counters is the production of multiple discharges, a spurious discharge being likely to occur immediately after any discharge. One possible cause of this is that in the discharge, atoms are excited into metastable states. A short time later these return to normal with the emission of radiation which may liberate photoelectrons from the wall of the counter, and thus initiate another discharge. Duffendack, Lifschutz and Slawsky[18] have advocated the use of pure hydrogen as the gas, since this is free from metastable states. Alternatively, a second gas may be mixed with the first to bring about de-excitation by collisions of the second kind, in which the energy of excitation is liberated as kinetic energy of the colliding molecules.

In the present state of knowledge of the factors affecting the initiation of the discharge each counter requires individual experimental investigation; nothing should be taken for granted. In particular, no assumption should be made as to the effective counting volume. Nor should the counter be assumed to retain constant characteristics for any great length of time. It is even necessary to check at intervals that the counter is able to add up correctly at a given operating voltage and temperature when connected to a given circuit. That is to say, counts should first

GEIGER-MÜLLER TUBE COUNTERS 101

be taken with no source of ionization, then with one source, then with another and then with both together. If this test is passed satisfactorily, it is not to be assumed that it will still hold if the operating voltage, temperature, or circuit is in any way altered; nor will it necessarily hold for a later time, the time interval being comparable with the time since the counter was made or previously tested. If there is the possibility of there being anything loose inside care should be taken in moving the counter, as it is possible that if the wire is not taut, the behaviour of the counter may change when it is turned over. It may also be noted that when glass chips were deliberately sealed up in a counter and the counter was shaken violently, it was found to acquire a high "natural" which decayed within a period of a few minutes. This effect was hypothetically ascribed to the re-formation of the oxide layer on the surface of the counter wall. It is to be remembered that all metal surfaces, including gold and stainless steel, after exposure to the air are covered by an oxide layer.

It should be noted that the addition test is also a test of the recording counting mechanism.

For satisfactory experiments with Geiger-Müller counters it is usually arranged that the results would not be affected by slight changes in the characteristics of the counters or by a certain number of spurious discharges. Experiments in which only coincident discharges in two or more counters are recorded are ideal in this respect, as the chance of spurious discharges coinciding with discharges in other counters is usually very small. In such experiments elaborate checks of the individual counters is generally unnecessary.

Growth of the discharge.

The first stage in the growth of the discharge is explained as an avalanche process. An electron is attracted towards the positively charged wire, and over a small region of its path it is accelerated between its collisions with atomic electron shells through more than the ionization potential of the gas. It is thus able to release a second electron from the next atom encountered. The two electrons are again accelerated and produce further ionization. The process may continue until a large charge is passing. It is found, however, that the charge passed is much greater than can be attributed to a single avalanche. Some process must therefore occur by which further electrons are liberated at points sufficiently far from the wire to initiate further avalanches. Two mechanisms for this process have been suggested and discussed (16,24,47); these may be designated the photoelectric and the positive-ion mechanisms. It is possible that both are simultaneously effective. Other mechanisms may also be imagined.

On the photoelectric hypothesis the radiation arising from the recombination of ions liberates photoelectrons from the wall of the counter. This hypothesis finds considerable support from the observed effect of the wall material on the form of the discharge.

On the positive-ion hypothesis a collision experienced by the positive ion is assumed to have a certain small probability of liberating an electron. The collision may occur either in the gas or at the wall. Some uncertainty exists about the quantitative aspect of this process; it is, however, to be noted that the magnitude of the photoelectric process is only known very roughly, though it appears to be of the order of magnitude required to explain the observed phenomena.

On both hypotheses the chance of maintenance of the discharge is approximately proportional to the number of positive ions present. The discharge therefore grows indefinitely but, if the maintenance process is sufficiently improbable, fluctuations are to be expected in the mean current passing.

Extinction of the discharge.

In order that the discharge shall be extinguished the probability of the maintenance process must be progressively reduced, and a number of mechanisms have been adopted to secure this. First, the voltage applied to the counter may be progressively reduced by some means depending on the discharge current; this reduces the number of electrons and positive ions produced in each avalanche. Two modes of extinction, which will be distinguished as resistance extinction and external extinction, operate in this way. In a third method of extinction the discharge is extinguished entirely by internal action.

Internal extinction.

The process of internal extinction depends on the low mobility of the positive ions. The electrons liberated in the avalanche process are quickly removed by passing to the wire. The positive ions, however, move more slowly, so that an excess of positive ions is left. There exists close to the external tube a potential gradient due to this distribution of positive space charge. Assuming that the electric field remains everywhere radial, this potential gradient at a radius a may be written as

$$-\frac{\partial V}{\partial r} = \frac{4\pi}{2\pi a}\int_0^a 2\pi r \, dr \, \rho$$

by Gauss's law, where ρ is the density of positive charge. If it is assumed that the potential difference between the wire and the external tube, that is, the voltage externally maintained across the tube, is either reduced or unchanged, then, to compensate for this added potential gradient close to the tube, the potential gradient must be reduced elsewhere. In fact, the electric field must be reduced close to the wire. The avalanche process is thus checked. If this process is to be effective it is necessary that the positive ions, at least near the tube, shall not be such that they are able to liberate electrons by any mechanism. The nature of the positive ion and the photoelectric character of the wall material is therefore again important.

It is found that this process of internal extinction is satisfactory and reliable if the vapour of some polar organic substance such as alcohol or acetone is included in the gas (74). Presumably the positive ions in such a mixture are rendered slow and have low recombination potentials.

Satisfactory internally extinguished counters may be constructed by using a very thin wire (about 50μ diameter) and not too low a gas pressure; 5–12 cm. of mercury is common. The inert gases argon or helium are particularly satisfactory. The pressure of alcohol vapour may be about 1 or 2 cm. Such counters have also been made using acetone vapour with no added gas. If the counter is to be permanent, wax, ebonite, and similar substances which absorb the vapour must not be used for the insulation of the wire from the tube. Glass or quartz may be used.

In order to detect the discharge it is usual to include a resistance of the order of 1 megohm in series with the counter. The voltage drop across this resistance when a discharge occurs plays no essential part in the extinction. Moreover, the duration of the voltage impulse across the resistance gives little indication of the time interval before the electric field within the counter is restored so that the counter is again sensitive.

With internal extinction the quantity of electricity passing in the discharge is approximately constant for a given operating voltage but increases rapidly with voltage over the normal operating range. If a large capacity is connected across the counter the voltage drop occurring with a discharge is reduced, as would be expected; the recovery time is not, however, necessarily lengthened, since the voltage drop may have been so small that the counter remained on the operating portion of its characteristic as far as the external circuit is concerned. The recovery time is conditioned by the time required for the removal of the space charge.

Resistance extinction.

The mode of action distinguished as resistance extinction is quite different from internal extinction, although in the operation of some counters both modes may occur simultaneously. The process of resistance extinction has been investigated in detail by Werner (77). He finds that in general there is a certain minimum current which may be passed in a continuous discharge through the counter; if the voltage across the counter is increased the current increases, but if an attempt is made to reduce the current below the minimum value by limiting the current or reducing the voltage then the discharge is extinguished. Let the voltage across the counter when the minimum current $i_{\text{min.}}$ passes be $V_{\text{min.}}$. If a higher voltage V_a is applied a discharge is not necessarily produced at once, but ionization within the counter initiates the discharge. If a resistance $R > \dfrac{V_a - V_{\text{min.}}}{i_{\text{min.}}}$ is included in series with the counter, then when the current $i_{\text{min.}}$ passes, the voltage across the counter will automatically be reduced to less than $V_{\text{min.}}$, so the discharge will not be maintained.

Although wide variations in $i_{\text{min.}}$ are possible, depending on the gas, the pressure, and the dimensions, in general if a voltage difference $V_a - V_{\text{min.}}$ of about 100 V. is to be permitted R will have to be about 10^8 to 10^9 ohms.

The process of extinction is again internal in the counter. As in other modes of operation the discharge, after initiation by the avalanche process, is maintained by a process depending on the current density. This mechanism would lead to an indefinite growth of the current for any voltage greater than $V_{\text{min.}}$, so that in order to obtain a limiting current it is necessary to suppose that an internal space charge builds up which lowers the electric field. In this mode, however, this space charge does not lead to the extinction of the discharge; it is necessary also to reduce the voltage externally applied to the counter. It may be noted that the voltage will not fall below $V_{\text{min.}}$. After this the voltage rises again, but the rate of rise is limited by the capacity of the counter. Some ions remain in the counter, but the number of ions must be sufficiently small to render maintenance of the discharge improbable.

This mode of operating counters with a high series resistance is probably still the commonest, though it suffers from severe limitations in the maximum speed of counting. If the resistance R is 10^8 ohms and the capacity associated with the wire and the recording apparatus connected to it is $30\,\mu\mu\text{F.}$, then the time constant for the recovery of the voltage on the counter is 3×10^{-3} sec. A second discharge occurring within about 0·01 sec. of another would produce only a diminished impulse.

The current through the counter ceases while the voltage is

low; after this there is no reason why the resistance should be large. Neher and Harper [51] have pointed out the advantages of a circuit which had previously been applied by Wynn-Williams [79], in which there is effectively a high resistance in series with the counter while current greater than a certain minimum is passing, but only a low resistance as soon as the current falls below this. The recovery of voltage is therefore rapid.

With resistance extinction the voltage across the counter drops to a certain point a little above the lowest operating voltage; the magnitude of the voltage drop is approximately proportional to the excess of the operating potential over this voltage and is independent of the capacity across the counter. This condition is rarely perfectly satisfied, since the process of internal extinction is usually operative to some degree.

Duration of discharge.

It has been pointed out that the mechanism of internal extinction prevents determination of the effective duration of a discharge from observation of the voltage change occurring across the counter. The duration can only be derived indirectly. If a recording counter is available with a resolving time less than the duration of the discharge, the minimum time between two discharges may be inferred from the results of the addition test at high rates of counting (see Chapter x). The duration is likely to be of the order of 10^{-4} sec., and a recording counter with a sufficiently short resolving time may not be available. In such a case a method based on coincidence counting may be adopted; each discharge renders the counter insensitive for the duration of the discharge, and the relative time during which the counter is insensitive may be inferred from the observed reduction of the number of true coincidence counts between it and another counter, when the counter under test is simultaneously exposed to a source of ionization giving a large but measurable rate of counting.

External extinction.

It will have been observed that in the previously described methods of extinction it is necessary that, in the process of removing the last of the ions from the counter, no free electrons should be released. If the choice of gas and wall material is restricted, it may happen that the liberation of free electrons is quite probable and prolonged discharges result. If the voltage across the counter is not restored until a predetermined time interval has elapsed, sufficient for the collection of all the ions, then such prolonged discharges are prevented. In a circuit devised by Wynn-Williams(79) this is achieved. For a counter to be suitable for external extinction in this manner it is only required that the minimum discharge current shall be sufficiently large to operate the extinguishing circuit. Very little systematic investigation has been made on externally extinguished counters. It may be possible to construct counters having valuable special properties which are only suitable for external extinction. For example, Wynn-Williams constructed counters which operate satisfactorily with an applied potential of only 300 V.

Circuits.

Three circuit arrangements which are used for resistance extinction are shown in Fig. 9·2. The first (*a*) requires no explanation; (*b*) and (*c*) differ in that in (*b*) the tube of the counter is at high potential with respect to earth, and in (*c*) it is at earth

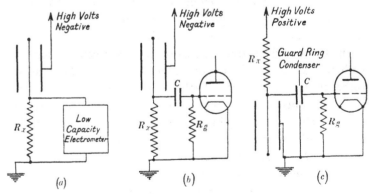

Fig. 9·2. Circuits for resistance extinguished counters.

R_x = extinguishing resistance 10^7 to 10^9 ohms.
R_g = grid leak resistance. $C = 2$ to $20\,\mu\mu\text{F}$.

potential. When, as is quite common, the tube of the counter is its outer wall, it is more convenient to be able to touch it without fear of shock, so circuit (c) would be adopted. This requires a condenser of rather special construction, as the insulation of the plate connected to the wire of the counter must be high compared with the counter series resistance, which may be 10^9 ohms. Further, if the condenser is simple the insulation between the plates must be at least 1000 times the resistance in the grid circuit of the valve if the high potential is not to alter the potential of the grid. This necessity is avoided if the insulation of the condenser is split with an earthed guard ring. The small condensers known commercially as ceramic condensers and rated for a working voltage of 750 usually have an insulation resistance greater than 10^{10} ohms and are often satisfactory if kept clean and not fingered. If the series resistance is 10^8 ohms or greater, the wire of the counter and the grid coupling condenser may also be touched without fear of shock.

Valve amplifying circuits for resistance extinction are shown in Fig. 9·3. In Fig. 9·3 (a) the grid of the valve is biased so that the valve is inoperative until the discharge current passing through R_X is sufficient to develop at least an appreciable fraction of a volt. When the valve is thus made to pass anode current the potential drop across the anode resistance R_A is added to that across R_X and the voltage on the counter is thus reduced. It should be noted that the current with which the discharge was formed was derived from the capacity of the counter together with that added by the circuit. When the voltage becomes insufficient to maintain the discharge this current rapidly falls, and when it becomes less than that flowing through the resistance R_X the potential rises again. Up to this stage the effective resistance in series with the counter may be shown by analysis to be M times R_X, where M is the magnification of the valve circuit. However, when the current through the counter ceases the valve becomes inoperative and the recovery time constant is that appropriate to the low resistance R_X.

Fig. 9·3 (b) shows a modification of this circuit in which the potentials applied to the valve are not above the normal rating. The vacuum in an ordinary valve used in the circuit of Fig. 9·3 (*a*) is liable to soften, due to the high anode voltage when the current is small.

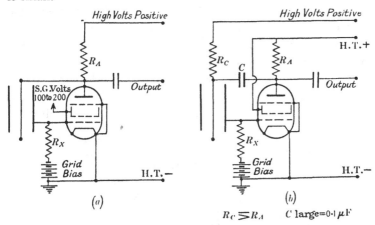

Fig. 9·3. Valve amplifying circuits for resistance extinction.

It should be particularly noted that it is the non-linear characteristic of the valve which gives this circuit an advantage over simple resistance extinction. Any modification of the circuit in which the non-linear property of the valve is not used must necessarily be equivalent to simple resistance extinction.

External extinction circuits.

One of the simplest circuits for external extinction is that of Wynn-Williams (Fig. 9·4). This is a complete recording counter circuit. The recorder is operated by the anode current of the thyratron. The armature of the recorder at the end of its travel breaks a contact which interrupts the anode circuit and extinguishes the arc in the thyratron. The armature is thus released and the contact is made again, thus restoring the operating potentials to the thyratron and Geiger-Müller counter. The extra battery for the Geiger-Müller counter may be replaced by a supply from a rectifier connected as in Fig. 9·3(*b*). In this form the counter is not suitable for very rapid counting, due to the inevitable slowness of the mechanical recorder.

The multivibrator circuit has been employed for external extinction (26,48,59). A successful circuit devised by de Sousa Santos (unpublished) is simply that of a multivibrator with the wire of the counter connected to the anode of one of the valves which is normally held in a non-conducting state by a high negative grid bias. A modification of this circuit in which the full Geiger counter potential is not applied to the anode of the valve is shown in Fig. 9·5. As soon as a discharge passes in the counter the wire of the counter drops slightly in potential; this drop of

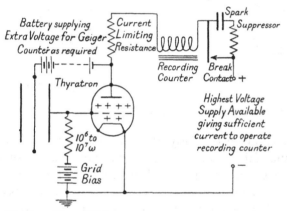

Fig. 9·4. Wynn-Williams external extinguishing circuit.

potential is amplified by the second valve and fed back to the grid of the first valve reversed in sign. If this impulse is as great as the applied negative cut-off bias the first valve passes current, its anode potential consequently falls and the process continues independent of what is occurring in the counter. The normal potential is restored again at a later time determined by the time constants of the coupling condensers and grid-leak resistances. Many other circuits using thyratrons, the multivibrator or a blocking oscillator would appear suitable.

Very little work appears to have been done on designing counter tubes most suitable for external extinction.

High counting rates.

The question of the minimum time for the discharge in a Geiger counter is important for a number of investigations.

Although the Neher and Harper circuit requires only a change of potential of the order of a volt on the grid of the valve, it nevertheless requires the passage of a considerable charge, since the capacity to earth of the case of the counter is usually several centimetres. The multivibrator circuit (Fig. 9·5) is free from this

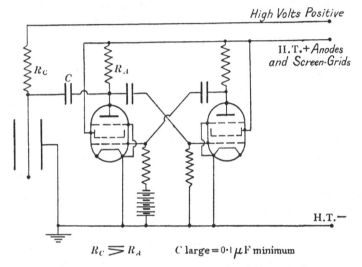

$$R_C \gtrless R_A \qquad C \text{ large} = 0·1\ \mu\text{F minimum}$$

Fig. 9·5. Multivibrator external extinguishing circuit.

limitation, since a change of potential on the wire of the counter of less than a tenth of a volt causes the potential to be removed from the counter in a time which may be less than a microsecond.

If only a single avalanche could produce sufficient potential drop to operate the recording circuit, the discharge time would be a minimum. This condition may be approached in two ways. If the processes responsible for the maintenance of the discharge can be suppressed, a sufficiently large avalanche may be produced with a very small chance of originating another and so building up the discharge. The change of potential produced by such an avalanche will be very small, but if such avalanches of 10,000 electrons can be produced it should certainly be possible to record them. The main difficulty in this line of approach is that the potential required on the counter is very critical.

The second line of approach is to employ an externally extinguishing circuit which operates on a single avalanche, to which the only restriction of size is that the potential on the counter shall not be so large that spontaneous discharges occur. The externally extinguishing circuit must operate in such a short time, and for such a time, that any subsequent avalanche originated by the first is so small that it would not contribute to the maintenance of a discharge.

Certain practical considerations should be noted. First, if advantage is to be taken of discharge times of the order of a microsecond the high-gain amplifier required must be designed to operate up to frequencies of the order of a megacycle. Such resistance-capacity coupled amplifiers are now technically possible and are employed in television systems. Secondly, it would be necessary to employ complete electromagnetic screening of the Geiger-Müller counter, such as is necessary for ionization chambers operating with linear amplifiers. Thirdly, the insulation of the Geiger-Müller counter would require the same consideration as that for ionization chambers.

It is not therefore suggested that the operation of Geiger counters in this simpler manner is any easier than the usual operation where discharge times of 10^{-4} sec. can be permitted, and screening and insulation difficulties are at a minimum.

It may be noted that the single avalanche is confined to a certain small region of the counter volume. The remainder of the counter will remain sensitive even while an avalanche is taking place. So, if the first method, of working without any externally extinguishing circuit, is adopted, the resolving time may be less even than the duration of a single avalanche.

Operated in these ways the Geiger counter is not necessarily less efficient than in the normal condition, for although the potential applied is less than that for the plateau in the normal regime, an avalanche must follow every ion pair produced except those very near the wire. Electrons produced far from the wire may also at high gas pressures be effectively lost by attachment to form molecular negative ions. The loss of efficiency at the low-voltage end of the plateau is, however, to be connected with a failure of the mechanism of growth and maintenance rather than the absence of avalanches.

The fundamental lower limit to the avalanche size required should be considered. If the single avalanche is to operate a valve amplifier, the limit is that imposed by thermal agitation noise and valve noise in the input stage of the amplifier. This has been discussed in Chapter III. It is common practice in amplifiers used for television and for the reproduction of sound from film, both of which are operated by photoelectric cells, to evade this thermal agitation and valve noise limit by amplifying directly the electrons liberated in the photocell either by ionization, by collision or now more commonly by secondary electron emission. The latter process is always adopted in television, where only very short time delays can be allowed. In the normal operation of the Geiger-Müller counter the process of ionization by collision is employed to amplify the discharge, and this introduces the inevitable time delay. It would seem therefore that if very short times are required the television technique of amplification by secondary emission might be employed either directly in the counter, in which case the best form of the counter would probably differ from that of the familiar wire along the axis of a cylindrical tube, or alternatively by registering the initial avalanche by the radiation which accompanies it with a photocell employing amplification by secondary emission.

For the successful operation of secondary emission multipliers with such very small currents it is necessary to ensure that the thermionic emission from the secondary emitting surfaces is very small. Z. Bay(5) has reported success with caesium electrodes cooled in liquid air. Also, for use at ordinary temperatures, he has found that surfaces coated with barium oxide activated in the manner of dull emitting cathodes give satisfactory multiplication by secondary emission with negligible thermionic emission. W. H. Rann(55) has also reported similar experiments with larger primary ionization currents.

Chapter X

STATISTICS OF RANDOM DISTRIBUTIONS

When the distribution in time of the particles arising from the decay of a long-lived radioactive substance is examined it is found to conform statistically with a purely random distribution. The statistics of such a random distribution are therefore important for the interpretation of counts of particles.

A purely random distribution would arise from the condition that the chance of a particle arriving in a small time dt is λdt, where λ is a constant provided that $\lambda dt \ll 1$. The arrival of a particle is independent of any other influence, in particular the chance is not influenced by the previous arrival of another particle or by the absence of a previous particle.

A definition of the chance of a particle arriving may be given as follows. If, in a very large number N of equal intervals dt, the number of particles counted is n, then λ is defined as such that $n = N\lambda dt$, when $n \to \infty$.

Let $W_n(t)$ be the chance that n particles and no more arrive in a time t. The value of $W_n(t)$ may be derived by the following process of mathematical induction due to Bateman(4). The chance that $n+1$ particles arrive in $t+dt$ is the sum of two chances: (1) that $n+1$ particles arrive in time t, and none in the extra time dt. This chance is $(1 - \lambda dt)W_{n+1}(t)$; (2) that n particles arrive in time t, and one in the extra time dt. This chance is $\lambda dt W_n(t)$. The chance that two or more particles arrive in the extra time dt $[(\lambda dt)^2 - (\lambda dt)^3$, etc.] becomes negligible as dt is reduced indefinitely. Hence

$$W_n(t+dt) = (1 - \lambda dt)W_{n+1}(t) + \lambda dt W_n(t)$$

or $\quad W_{n+1}(t+dt) - W_{n+1}(t) = \lambda dt[W_n(t) - W_{n+1}(t)].$

Proceeding to the limit

$$\frac{dW_{n+1}}{dt} = \lambda(W_n - W_{n+1}).$$

Putting $n = 0, 1, 2, \ldots$ in succession, we have

$$\frac{dW_0}{dt} = -\lambda W_0,$$

$$\frac{dW_1}{dt} = \lambda(W_0 - W_1),$$

$$\frac{dW_2}{dt} = \lambda(W_1 - W_2),$$

$$\ldots\ldots\ldots\ldots\ldots\ldots\ldots$$

These equations may be solved by multiplying each by $e^{\lambda t}$ and integrating. Then since $W_0(0) = 0$, we have in succession

$$W_0 = e^{-\lambda t}.$$

Then

$$dW_1/dt = \lambda e^{-\lambda t} - \lambda W_1, \quad \text{or} \quad e^{\lambda t}dW_1/dt + \lambda e^{\lambda t}W_1 = \lambda$$

or

$$\frac{d(W_1 e^{\lambda t})}{dt} = \lambda.$$

$$W_1 = \lambda t e^{-\lambda t},$$

$$W_2 = \frac{(\lambda t)^2}{2!} e^{-\lambda t},$$

$$\ldots\ldots\ldots\ldots\ldots\ldots\ldots$$

Hence

$$W_n = \frac{(\lambda t)^n}{n!} e^{-\lambda t}.$$

The average number of particles which arrive in time t is λt.

Write $\lambda t = x$. The chance that n particles arrive in t is $W_n = \dfrac{x^n}{n!} e^{-x}$.

This relation is known as Poisson's Law, and has been derived in other ways from the assumption of a random distribution in different forms [73].

The number of particles N counted in a large time T will in general not be exactly λT. The mean deviation D, defined as $\sqrt{[\text{probable value of } (N-\lambda T)^2]}$, is given by

$$D^2 = \sum_{N=0}^{\infty} (N-\lambda T)^2 \frac{(\lambda T)^N}{N!} e^{-\lambda T}$$

$$= e^{-\lambda T} \sum_{N=0}^{\infty} \left[\frac{(\lambda T)^N}{(N-2)!} + \frac{(\lambda T)^N}{(N-1)!} - 2\frac{(\lambda T)^{N+1}}{(N-1)!} + \frac{(\lambda T)^{N+2}}{N!} \right]$$

$$= e^{-\lambda T} \sum_{N=0}^{\infty} \frac{(\lambda T)^N}{(N-1)!} = \lambda T.$$

Hence $D = \sqrt{(\lambda T)}$.

The numerical calculation of Poisson's Law distributions may be carried out by observing that $W_n/W_{n-1} = x/n$. If the probable number is 10 and the chance of observing 10 is written as p,

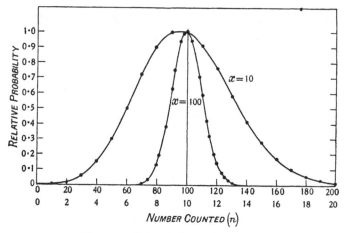

Fig. 10·1. Poisson's Law distributions.

then the chance of observing 9 is also p. The chances of observing 8, 7, 6, etc. are successively $0·9p$, $0·8 \times 0·9p = 0·72p$, $0·7 \times 0·72p = 0·504p$, etc. Poisson's Law distributions for $x = 10$ and $x = 100$ are shown in Fig. 10·1.

Loss at high rates of counting.

When particles arrive at random at a mean rate of p per sec. some will arrive too close together to be counted separately.

Suppose r per sec. are recorded and the minimum resolving time of the counter is $1/q$ sec. The counter is then effectively unable to operate for r/q sec./sec., during which time pr/q particles arrive and are not recorded. We therefore have

$$r = p - pr/q, \quad \text{or} \quad p(1 - r/q) = r, \quad p = r(1 - r/q)^{-1}. \quad (10\cdot1)$$

For example, suppose $q = 25$, i.e. the counter takes $\frac{1}{25}$ sec. to record, and suppose 25 particles are counted per minute, i.e. $r = \frac{25}{60}$, then $p = \frac{25}{60}(1 - \frac{1}{60})^{-1}$, or the true rate of arrival of particles is $1\cdot7\%$ greater than the measured rate. If 250 particles per minute were counted, $p = \frac{250}{60}(\frac{6}{5})$, i.e. 300 per min. It seldom happens with counting meters that the recording time is known or constant to better than 10 %, so an uncertainty of at least 2 % is added to this rate of 300 per min. This would in many cases be unimportant, but sometimes a greater certainty is required. A counter with a shorter resolving time must then be used.

If an electrical counter with a resolving time of $\frac{1}{5000}$ sec. is used the loss will amount to 2 % when counting 6000 per min. It may be pointed out as a general relation that if the resolving time is $1/q$ *second* the loss is approximately $1\cdot7\%$ when counting q impulses per *minute* and 2 % when counting $1\cdot2q$ impulses per minute.

The above simple considerations do not always apply, for it is to be noted that if the ionization chamber or discharge counter is unable to produce two impulses within a time less than $1/q$ sec. the recording counter introduces *no loss at all*, and the correction to obtain the true rate of arrival of particles must be determined from the characteristics of the ionization chamber and its amplifier.

If, on the other hand, the resolving time of the recording counter is slightly longer than that of the ionization chamber and its amplifier, at high rates of counting both resolving times must be taken into account. Let the resolving time of the ionization chamber and amplifier be $1/c$ sec. Again let r particles per sec. be counted, the recording counter is then dead for r/q sec./sec. After each recorded impulse the ionization chamber is dead for $1/c$ sec.; there is therefore a time $r/q - r/c$ sec./sec. during which the ionization chamber might produce an impulse which is not recorded. The occurrence of impulses within this time may extend the dead time of the ionization chamber beyond the limit of the dead time of the counter.

A rigid calculation of the loss to be expected is complicated, but if the simplifying assumption is made that each particle produces a dead time of $1/c$ sec., as far as the recording counter is concerned, no matter whether it arrives within $1/c$ sec. after another or not, an upper limit to the correction may be obtained as follows. After each recorded impulse the recording counter is dead for $1/q$ sec., and any particle arriving in the last $1/c$ sec. of this interval will extend the dead time on the average by $1/2c$ sec., assuming the chance of two particles arriving in the interval to be negligible. Thus we have the number of particles arriving in r intervals of $1/c$ sec. $= pr/c$. Each provides an extra dead time $= 1/2c$ sec., therefore the extra dead time $= pr/2c^2$; during this time $p^2r/2c^2$ particles arrive and are missed. The number counted is thus

$$r = p - pr/q - p^2r/2c^2$$

$$= p\left[1 - \frac{r}{q}(1 - pq/2c^2)\right].$$

Suppose $q = 100$, $c = 200$, $r = 20$ per sec. $= 1200$ per min. Then

$$p \simeq r[1 - \tfrac{1}{5}(1 + \tfrac{1}{32})]^{-1} = \tfrac{160}{127}r = 1512 \text{ per min.}$$

Without the extra correction p is found as 1500 per min. The number 1512 is, however, an upper limit because the time added for two particles arriving in any of the intervals of $1/c$ sec. has been $2 \times 1/2c$ sec., and this is too great; also it is unlikely that the ionization chamber arrangements will be such that every particle will produce a dead time $1/c$ sec. whether it produces an impulse or not.

This example indicates that great care is necessary in estimating corrections in such complicated circumstances. One of the commoner pitfalls is to write p' = number of impulses per sec. from the ionization chamber and then apply equation (10·1) with p' substituted for p to obtain p' from r the number counted.

This is not admissible because equation (10·1) assumes that the p particles are at random; the p' impulses from the ionization chamber are certainly not at random, since no intervals shorter than $1/c$ sec. can occur.

Caution should be observed in applying any large correction to a number counted, for many complicating circumstances may exist. If the rate of arrival of particles is very great the recording counter may cease to operate at all, it may record steadily at as fast a rate as possible, or it may record steadily as fast as possible up to a limit at which the ionization chamber and amplifier fail to produce large enough impulses. For such and higher rates few or no impulses are recorded. Each of these conditions requires a different correction when counting at high speeds.

A complicated correction is sometimes necessary with a scale-of-two counter. If the counting meter operates for every eighth impulse and cannot operate twice in less than $1/m$ sec. there is a chance of loss if nine impulses occur within $1/m$ sec. This occurs when 8 particles follow a certain particle which trips the counting meter ($\frac{1}{8}$ of all the particles are in this category) within $1/m$ sec. Since 8 particles are lost each time this occurs the fraction lost will be $8 \times \frac{1}{8}$ of the chance that 8 particles or more arrive in a given interval of $1/m$ sec., viz.

$$\frac{x^8 e^{-x}}{8!} + \frac{x^9 e^{-x}}{9!} + \frac{x^{10} e^{-x}}{10!} + \dots,$$

where $x = p/m$ and p is the mean number of particles arriving per sec. If, however, any of these 8, 9, 10, … particles arrive within $1/q$ sec. after another it will not be recorded by the first stage and will not therefore introduce loss by the counting meter. This chance is $n^2 m/q$, where $n = 8, 9, 10, \dots$. The nett loss by the counting meter is therefore

$$\frac{x^8 e^{-x}}{8!}\left(1 - \frac{64m}{q}\right) + \frac{x^9 e^{-x}}{9!}\left(1 - \frac{81m}{q}\right) + \frac{x^{10} e^{-x}}{10!}\left(1 - \frac{100m}{q}\right) + \dots.$$

This must be added to the loss introduced by the resolving time of the first stage of the counter. In normal practice the loss by the other stages of the counter is negligible. A table is reproduced below giving some illustrative figures for the special case where the resolving time of the first stage is $\frac{1}{5000}$ sec. and the resolving time of the counting meter is $\frac{1}{25}$ sec.

Mean rate no./min.	Loss by first pair %	Loss by meter %	Nett loss %
1500	0·5	7×10^{-4}	0·5
3000	1·0	0·07	1·1
4500	1·5	0·75	2·3
6000	2·0	3·2	5·2
7500	2·5	8·0	10·5

It should be pointed out that the above calculation is only approximate, and applies only when the losses are so small that in the correction terms it is unnecessary to distinguish between p, the mean rate of arrival of particles, and r, the mean rate of impulses passed by the first pair.

Chapter XI

COINCIDENCE COUNTING

Coincidence counting has already been mentioned in the discussion of Geiger-Müller counters in Chapter IX. The very simple and highly satisfactory mixing circuit which is commonly employed has also been discussed in Chapter VII dealing with non-linear properties of valve circuits. The simplest coincidence counting system consists of two internally extinguished or resistance-extinguished Geiger-Müller counters feeding direct into such a mixing circuit. For certain purposes a greater refinement is necessary and the whole technique must be discussed in greater detail.

A swift ionizing particle such as might produce discharges in two counters would have a velocity of at least about 10^9 cm./sec. If the counters are less than a metre apart the ionization will be produced in both within 10^{-7} sec. of each other. Such a small time interval may generally be neglected. It so happens, however, that the initiation of the discharge in a Geiger-Müller counter, or the separation of the ions in an ionization chamber, occupies a finite time, and it is found that we must allow impulses to be regarded as coincident if they are timed within an interval which, according to circumstances, amounts usually to one or a few microseconds.

Suppose the duration of a discharge is 10^{-4} sec., then the two impulses might appear as in Fig. 11·1, in which the time scale is very large. Such impulses must be counted as coincident, but if two impulses produced by independent ionizing particles occur as close together as shown in Fig. 11·2, they should not be recorded as coincident.

It is quite practicable to make a coincidence-counting system capable of distinguishing between these. It would not be practicable to do this with the simple mixing circuit already described.

If the size of the impulse on each grid were carefully regulated it might be arranged that the valve was only rendered non-conducting by an impulse exceeding say V_b (Figs. 11·1, 11·2), but it

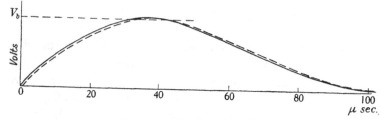

Fig. 11·1. Coincident impulses.

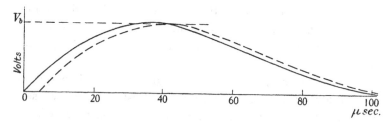

Fig. 11·2. Close impulses.

will be seen from Fig. 11·2 that V_b would have to be very close to the peak of the impulse if the two impulses are not to exceed V_b coincidently.

The usual method of solving the problem is therefore to differentiate the impulse with respect to time. This may be done by a simple resistance-capacity filter (Fig. 11·3) in which the time con-

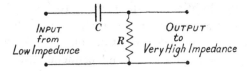

Fig. 11·3. Simple resistance-capacity filter.

stant CR is made short compared with the required resolving time. After passing through such a filter the impulses of Fig. 11·2 would appear as in Fig. 11·4, the ordinates in which are proportional to the slope of the curve of Fig. 11·2 averaged over a time of about a microsecond. These impulses may with advantage again be differentiated, producing impulses such as are shown in Fig. 11·5.

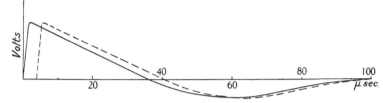

Fig. 11·4. Impulses of Fig. 11·2 after differentiation.

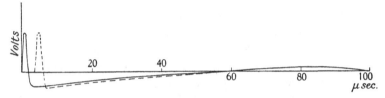

Fig. 11·5. Impulses of Fig. 11·4 after differentiation.

Each process of differentiation reduces the size of the impulse, so that it is usually necessary to amplify after each differentiation.

The impulses shown in Fig. 11·5 may be directly applied to the mixing circuit. In practice, however, these twice differentiated impulses are liable to vary considerably in size, so they may with advantage be passed through a simple discriminator stage (Chapter VII) to even up the impulses and discriminate against the subsidiary peak which occurs 80 μ-sec. after the initiation of the discharge (Fig. 11·5). It would even be possible to arrange a discriminator to pass on a very short pulse just at the moment when the impulse of Fig. 11·5 reaches its maximum, but this has probably never been found necessary.

It should be remembered that, if the resolving time is of the order of 10^{-6} sec., the amplifier must amplify at frequencies from 10^4 or less to over 10^6 c./sec.; the amplifier should therefore be regarded as a radio-frequency wide-band amplifier in which the layout of the wiring is as important as the circuit diagram.

The argument advanced in discussing the linear amplification of ionization impulses that only one quick coupling-stage, and that of not too short a time constant, may be used does not apply in the present considerations as it is not required that the voltage of the final impulse shall be proportional to that of the input impulse. In fact, by these processes of successive differentiation, the magnitude of the final impulse is proportional to the maximum values of dV/dt, d^2V/dt^2, d^3V/dt^3, etc., where V is a function of time representing the voltage of the initial impulse. The reduction of the resolving time for coincidences depends on the fact that these higher differential coefficients have their maxima very close to the moment of initiation of the discharge.

With this knowledge of what is technically possible consideration may be given to the conditions which it is likely to be necessary to satisfy in the coincidence counting of β-rays and γ-rays.

Probably the chief reason for adopting coincidence counting is to reduce the ratio of the natural effect to the effect under observation. When a Geiger-Müller counter with thin walls is unshielded it will give about 2 impulses per minute per cm.² of its maximum projected area. This is due to cosmic rays and ordinary radioactive contamination. By surrounding the counter with a thickness of 3 or 4 cm. of lead on all sides, the lead being free from radioactive contamination, this number of impulses will be reduced to about 0·5 per cm.² per min. With an average counter 10 cm. long by 2 cm. diameter, this amounts to about 10 impulses per minute.

The efficiency of a Geiger-Müller counter for recording β-rays is approximately unity, so that if a β-ray traverses two counters the probability of coincident discharges is very high. The number of coincidences would therefore be only slightly less than the number of single counts. The β-rays must, however, be limited to directions which traverse both counters. The ionization produced by cosmic rays is largely due to single tracks having an ionization density equal to or greater than that of β-ray tracks, but a majority of these have their directions making an angle of less than 45° with the vertical. By placing two counters with their axes parallel and horizontal just so far apart that only rays making angles of less than 30° with the horizontal can traverse both counters the number of coincident discharges due to cosmic rays will be about one in 4 min., with the average size of counter, 8 cm. long × 2 cm. diam.

In addition to this "natural" count of true coincidences, other coincidences will occur due to two independent discharges happening by chance within the time interval set as the limit of resolution. This time interval may be reduced to a few microseconds, but nevertheless, if the rate of discharges in the individual counters is high, these chance coincidences may outnumber the coincidences which are being studied. It would be impossible to give a general discussion covering all cases and the experimenter must be left responsible for investigating all sources of coincidences effective in each experiment.

One paradoxical effect may however be noted. If other things remain constant a reduction of the duration of the pulse in a counter is liable to lead to an increase in the number of chance coincidences. The number of chance coincidences N_c per sec. for a resolving time interval for coincidences of τ sec. is $2\tau N_1 N_2$, where N_1, N_2 are the numbers of discharges per second in the two counters. The factor 2 arises because the discharge in the second counter may be either before or after that in the first by an interval up to τ sec. If the duration of the pulses is reduced N_1 and N_2 will be increased for a given source of ionization (see Chapter x).

The formula $N_c = 2\tau N_1 N_2$ is almost self-evident if the N_1 and N_2 discharges occupy a negligible fraction of the total counting time, but the formula is not valid if this is not the case.

Chapter XII

ENERGY DETERMINATIONS FROM RANGE MEASUREMENTS

The theory of the loss of energy of charged particles by the ionization processes has now been brought into fairly satisfactory agreement with the experimental observations [7,9,19]. The theory for heavy particles may be summed up in the proportionality

$$\frac{dT}{dx} \propto N \frac{Z^2}{v^2} f\left(\frac{v^2}{E}\right),$$

where T, Z and v are the kinetic energy, charge number and velocity of the projectile. N is the number of atoms per unit volume. $f(v^2/E)$ is a numerical function of v^2/E, where E approximately represents the energy loss per ion pair produced. Thus for different particles having the same *velocity* the specific ionization is proportional to Z^2. For heavy particles for which $T \simeq \frac{1}{2}Mv^2$ (where M = mass of the projectile) having a given velocity v it follows that the total range R is proportional to M/Z^2.

It is thus seen that the form of the specific ionization-range curves, such as are shown in Fig. 1·1 for different particles, are of approximately the same shape but with the scale of ranges proportional to M/Z^2 and the ionization scale proportional to Z^2.

There is, however, a process which disturbs this simple relation. When the particle is travelling relatively slowly it is liable to capture an electron or even two if it is doubly charged; it then travels for a space with a reduced effective charge before again losing the captured electron. This phenomenon of capture and loss was studied for α-particles first by G. H. Henderson [30] and then in greater detail by Rutherford [60]; a full summary of these and later investigations is given in *Radiations from Radioactive Substances* [61]. The number of interchanges of charge along the track was found to be very great, amounting to more than a thousand. Except near the end of the range the mean free path for loss is much smaller than the mean free path for capture, so the particle carries a single charge for a relatively short part of its range, but in the last 4 mm. of its range in standard air the particle is more often singly charged than doubly. It was found that the equilibrium ratio between singly and doubly charged particles, which is a measure of the ratio of the mean free path for loss to the mean free path for capture, for particles of a given velocity was almost independent of the atomic weight of the absorbing substance, the only marked departure being for high velocities in hydrogen.

It must be pointed out that considerable uncertainty exists about the form of the relation connecting the specific ionization and the range for particles of low energy both theoretically and experimentally(78). The related function connecting the rate of energy loss and the range is similarly uncertain, and also the range integral of this, which is the energy-range relation. Since it is convenient to make measurements of ranges in order to deter-mine energies, the uncertainty about the range-energy relation for very short ranges is unfortunate. Means were, however, devised by which energies might be determined from range measurements in such a way that this uncertainty is immaterial. While it is easy to measure the range of a homogeneous group of α-particles approximately, precise measurements are very diffi-cult and the probable error of the best determinations is 0·2 mm. of standard air or more. On the other hand, the differences between the mean ranges of a number of homogeneous α-particle groups from radioactive substances have been measured with a probable error of 0·03 mm., which represents an accuracy of about 2000 e.V. in the corresponding energy. It has therefore been necessary to fix a conventional scale of ranges to which measurements may be referred. This convention also requires an exact definition of the conditions of measurement.

To emphasize the distinction between this conventional scale of ranges and the absolute scale it should be pointed out that the energies of α-particle groups may be deduced from measurements made on the conventional scale to about 2000 e.V. between about

5 and 10 M.e.V. Measurements attempted on the absolute scale at present diverge by as much as 1 mm. of standard air or 100,000 e.V.

The techniques of measuring α-particle ranges and energies depend on the fact that radioactive substances can be deposited on surfaces in such a way that few of the emergent α-particles lose any appreciable energy before escaping from the surface. The energies of the α-particle groups from such sources have been determined with great accuracy.

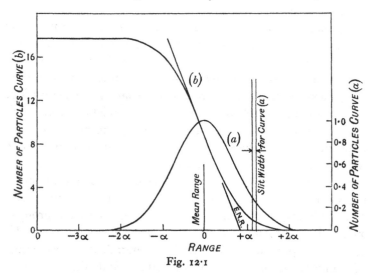

Fig. 12·1

The standard absolute energy determination has been made by Briggs (14), who measured by the method of magnetic deflexion the energy of the group of α-particles from radium C'. The absolute energy of α-particles in this group was found to be $7\cdot6802 \times 10^6$ e.V., with a probable error of only 7 in 10^5, i.e. 540 e.V.

This is actually determined from the measured $H\rho$ (ρ = radius of curvature of path in magnetic field H) of the particle and the value of E/M for the α-particle which is determined from the Faraday and the ratio of M for the α-particle to the chemical unit atomic mass. The measurement of H and the Faraday involves the relation between the absolute and international electrical units. There is at present a discrepancy of 8 or 9 in 10^5 between the determinations of this ratio at the different standardizing laboratories. This accounts for part of the probable error in the energy of the α-particle.

The relative velocities of the α-particles from a number of radioactive substances have been determined with varying accuracy from 1 in 10^5 = 70 e.V. to 1 in 10^3 = 7000 e.V. by Briggs (13), Rosenblum (58), and Rutherford, Wynn-Williams, Lewis and Bowden (39, 64).

The definitions of ranges on the conventional and absolute scales should therefore be made with sufficient precision to allow the determination of energies to about 1000 e.V., should it be possible to measure the ranges with the necessary accuracy.

The absolute range of a single particle is taken as the distance travelled measured along the track from the source to the last ion produced. The range of a particle of a given initial energy is subject to a variation due to the statistical nature of the ionization process. The average or mean range of a large number of particles initially all of the same energy will however preserve a precise meaning. The fact that the variation of the individual ranges or straggling has a definite form enables other ranges to be exactly defined. Two such other ranges have been extensively used, these may be distinguished as the extrapolated ionization range and the extrapolated numbers range.

The distribution of ranges has the symmetrical Gaussian form Fig. 12·1 (a). This may be deduced from the well-known argument that if after any given energy loss the distribution of distances travelled is Gaussian, then since the probable distance of travel depends only on the energy of the particle this Gaussian form will be conserved. If then after a large number of collisions the inhomogeneity is still small compared with the spread finally produced, so that a negligible error would be introduced by assuming the distribution at this stage to have the Gaussian form, then the final distribution will conform very closely to the Gaussian form, which may be represented by $N_{R+r} = N_R e^{-r^2/\alpha^2}$; N_{R+r} is the number of particles with a range $R+r$, and R is the mean or average range. The quantity α will be referred to as the straggling parameter, following Livingston and Bethe (42); it is the same as $\rho_3 x$ introduced by Briggs (12) and the quantity ρ used by Rutherford, Ward and Wynn-Williams (63) and later writers in this group and called somewhat loosely the straggling "coefficient". The straggling parameter is usually expressed in millimetres of air. If the α-particle ranges have this Gaussian distribution, then by plotting the number of particles passing beyond a given range against the range a curve of the form of Fig. 12·1 (b) is obtained. The intersection of the tangent at the point of inflexion of this curve with the axis of zero number is the *extrapolated numbers range*. It is easy to calculate that this range exceeds the mean range by $\alpha \dfrac{\sqrt{\pi}}{2}$.

In the early work on α-particle ranges, extrapolated ranges having a different meaning were measured. This extrapolated range should be distinguished as the extrapolated ionization range and is obtained not by counting the α-particles but by measuring the ionization produced in a short distance by a pencil of the rays. This ionization plotted against the range is the well-known Bragg ionization curve. If the tangent at the point of inflexion is extended to cut the scale of ranges, the range indicated is the *extrapolated ionization range*. In practice it is usual to draw this tangent on the assumption which is approximately correct that the straight portion of this curve is long; this makes the range slightly longer.

For the purpose of measuring ranges standard air is taken to be dry air at 15° C. and 760 mm. pressure measured under standard barometric conditions, in which the mercury column is reduced to the equivalent height at 0° C. at sea level at 45° latitude. The magnitude this correction is likely to assume does not appear always to have been appreciated. If the mercury column of the barometer is at a temperature of 15° C., the equivalent height at 0° C. will be 0·25 % less, corresponding to a correction in a range of 10 cm. of 0·25 mm. or 15,000 e.V. The height and latitude correction in Cambridge amounts to only 0·066 %, but may be much larger in other laboratories.

The approximate range in any medium other than air may be obtained by dividing the range in standard air by an appropriate factor termed the relative stopping power of the substance. For example, an average value for the stopping power of water vapour relative to that of air at the same temperature and pressure may be taken as 0·74. Typical laboratory air in England may be taken to have a humidity of 75 % at 16° C. The corresponding water-vapour pressure is 1·0 cm. Very considerable departures from this, however, are likely to be encountered. For this example the correction to dry air would be about 0·34 % or 20,000 e.V. for 10 cm. range.

Stopping screens.

It is common practice to use uniform thin sheets of mica to reduce the range of particles, and the distance by which the range is thus reduced is then the "stopping power" of the mica sheet. Caution is necessary in applying this conception of stopping power, for accurate measurements show that it depends on the initial velocity of the α-particle, and this variation is very great if the atomic weight of the substance is very different from that of air. The general results obtained by Gurney (28) and by Marsden and Richardson (45) are shown in Fig. 12·2.

A mica sheet used in range measurements should be placed so that the emergent range of the α-particles is the same as used in the calibration of the sheet.

It is convenient in the laboratory to calculate the stopping power of a mica sheet from its mass per cm.² or from its thickness. Recent experiments by Bennett (6) have shown that this procedure should not be relied on to an accuracy better than 2 or 3 %, even for pieces of mica from the same batch. If, however, the mica is thin, say of less than a centimetre stopping power, such uncertainty is often negligible. It is by no means easy accurately to measure the stopping power of a mica sheet. For mica known as Green Madras the stopping power of a sheet weighing

1·43 mg./cm.^{2} is 1 cm. of standard air with a probable uncertainty of 1%, and this applies only for α-particles having a residual range after leaving the mica between 3·5 and 10 cm.

When mica is split for this work it should be very carefully examined in reflected *monochromatic* light to ensure that the

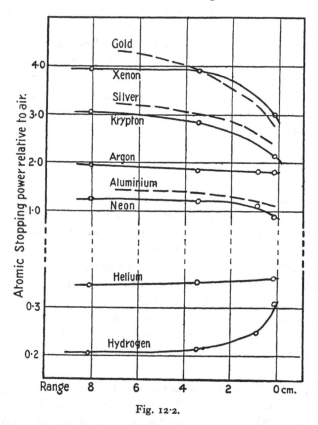

Fig. 12·2.

thickness is uniform. Non-uniformity shows up as a break in the intensity of the reflected light caused by the change in the interference between the light reflected at the two surfaces. These changes of thickness may be very slight, amounting perhaps to only $\frac{1}{10}$ of a wave-length, so very careful examination from all angles is called for.

An optical lever device for measuring the thickness of mica sheets is found very useful.

When protons are being observed mica-stopping screens are generally to be avoided, because mica contains hydrogen which may be projected as fast protons by other protons, α-particles, neutrons or other heavy particles. Such protons knocked on by α-particles may have a range four times that of the α-particles, but are distinguished from the α-particles by their lower specific ionization. The number of protons varies greatly with the angle at which they are knocked on, but as a rough order of magnitude if one proton for 10^4 α-particles would be undesirable, an estimate of the number should be made though it may be found that there are 100 times fewer. The relevant information may be found in *Radiations from Radioactive Substances*, pp. 252 *et seq.* [61]. Where knocked-on protons are undesirable thin sheets of aluminium may be used as stopping screens; the uniformity of such screens is, however, difficult to test.

Finally, in the measurement of ranges well defined and carefully planned geometrical conditions are essential.

The use of mica or aluminium absorbing screens enables the ionization chamber to be brought much closer to the source. Thus the ionization chamber subtends a greater solid angle at the source and more particles are received. Some of the tracks of these particles will consequently make an appreciable angle (θ) with the principal direction, and their range will be $\sec\theta$ times their apparent range. This "secant effect" may be considerable if the range is long. For example, $\sec 10° = 1\cdot015$, making a change of 8 mm. in a 50 cm. range which is not uncommon for protons.

Absolute measurements of range may be attempted directly in the Wilson expansion chamber. In practice it has proved difficult to measure ranges with great accuracy by this method; there is always the uncertainty of the exact composition, density and temperature of the gas and vapour mixture. Then, owing to the phenomenon of straggling it is necessary to measure some thousands of tracks to eliminate error from the statistical variations of the mean range.

The old method of measuring the extrapolated ionization ranges gives consistent results more readily, but the exact interpretation of the results depends on the form of the specific ionization-range relation for a single particle near the end of its range.

In the counting method particles giving more than a certain ionization beyond a certain range from the source are counted. This method enables differences of mean ranges or extrapolated number ranges to be determined with accuracy, but again the uncertainty of the specific ionization-range relation near the end of the range prevents the deduction of the exact absolute range.

The conventional range scale.

On the conventional scale of ranges, the mean range of the α-particles from thorium C′ is taken to be 8·533 cm. in standard air, and other ranges are to be measured by differences from this in standard air.*

It is of interest to consider how this conventional scale was fixed. The only method which gives the absolute range independent of the form of the specific ionization-range relation near the end of the range is that of direct measurement in the Wilson expansion chamber. The conventional range scale should therefore be made to agree as closely as possible with what would be determined by expansion-chamber measurements. There are, however, not many accurate range determinations made in expansion chambers available. Measurements of extrapolated ionization ranges may be related to expansion-chamber ranges most directly by using an ionization-range relation determined in expansion-chamber measurements.

Combining the ionization-range relation determined by Feather and Nimmo [22] from expansion-chamber measurements with the appropriate straggling coefficients determined from counting measurements, it is possible by numerical calculation to find the difference between the extrapolated ionization range

* Holloway and Livingston [32] have recently proposed a conventional scale making the thorium C′ range 8·570 cm.

and the mean range. This was found to be $0.8\alpha - 0.06$ mm. for ranges of α-particles between 3 and 12 cm., where α is the straggling parameter in mm. It should be understood that this value may have little absolute significance, but it is only required in order to relate range measurements made in the expansion chamber to determinations of extrapolated ionization ranges. The mean range is the average range to the last drop condensed on the track in an expansion chamber. Other determinations of the ionization-range relation at short ranges have been reported made by electrical measurements (32,35,65,70,76). These results differ considerably, but they have no importance in this connexion as they necessarily lack any point which can be identified as the point at which the last drop is formed in an expansion chamber.

The best expansion-chamber measurements of mean ranges, and the mean ranges derived from measurements of the extrapolated ionization ranges by the relation mentioned above, were then compared with the differences between the mean ranges found by the counting method. Combining the data in this way it appeared that the expansion-chamber value of the mean range of the thorium C′ particles was best given as 8·533 cm. (40,62).

The absolute accuracy of this figure is of no importance for the experimental determination of energies from ranges measured on the conventional scale.

The standard determinations of the relative velocities of radioactive α-particle groups fix the α-particle conventional range-energy relation down to the range of polonium 3·805 cm. ($5·30 \times 10^6$ e.V.). This has been extended to lower energies mainly by the experiments of Mano (44). Theoretical extrapolations have also been made by Duncanson (19) and by Bethe (42). The relation must, however, be considered rather less certain in this low-energy region than in the range covered by the standard α-particle groups.

The most accurate measurements of ranges by the counting method have been made with a differential chamber, which, as explained in Chapter II, enables counts to be made only of particles stopping in the chamber within narrow limits of range which may be as little as 2 mm. This narrow band is referred to as the effective slit width. The variation of the form of the number-range curve as this effective slit width is altered is illustrated in Fig. 12·3 for a homogeneous group of α-particles.

A common defect of radioactive sources is that they are dirty, that is to say there is some foreign matter covering the radioactive material. Some particles therefore leave the surface with a reduced velocity due to their passage through this layer of matter. The result of this is that the tail on the short-range side of a group

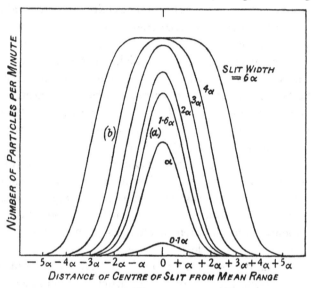

Fig. 12·3

is increased. Allowance must be made for this in fitting a curve of the correct shape for a clean source to the experimental points. If the dirtiness is considerable it is difficult to know how to fit the curve, and in such cases it is found that the apparent extrapolated numbers range is much less reduced than the apparent mean range.

When the α-particles originate from a process of artificial transmutation it often happens that they are liberated at a considerable depth below the surface. If we suppose that the particles are liberated uniformly throughout the material of the target the resulting numbers-range curve (counting all particles exceeding the range) is as shown in Fig. 12·4. It might be supposed that extrapolation of the straight portion of this curve would

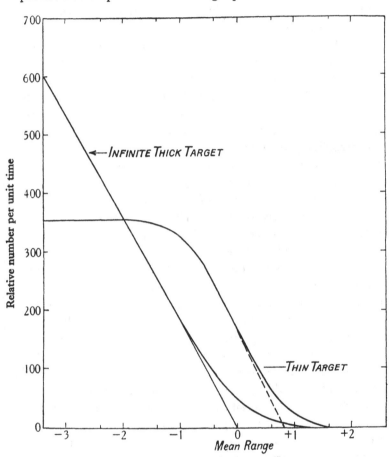

Range in units of straggling parameter

Fig. 12·4

lead to the extrapolated numbers range, but that this is not the case is evident from the other curve in Fig. 12·4, which shows the form of the numbers-range curve obtained from a thin target. No confusion between these two curves should arise for the straggling parameter, for a given range is a definite quantity and this accounts for the whole breadth of the sloping portion of the thin-target curve. If the sloping portion is seen to extend over a greater range than this, either the target must be thick or the emitted particles are not homogeneous in velocity. This is of course to be expected in artificial transmutations if particles emerging at different angles with the incident beam are included in the observations. Interpretation is only difficult if both the target is thick and the particles are emitted with a continuous distribution of velocities of unknown form. It should be noted that a source is effectively thick if its thickness is comparable with the straggling parameter of the particles under observation. For this purpose the straggling parameter may be estimated as a little greater than 1 % of the range for α-particles and 2 % for protons. It proves in practice difficult to be certain that the exact thickness of a source is known. The roughness of the surface must be taken into account. The chance of scattering in the source itself is also sometimes to be considered.

When α-ray tracks are observed in a Wilson expansion chamber it is found that the ends of individual tracks are very often bent; the range measured along the track is therefore not quite the same as the range measured as the maximum distance from the source reached by the particle. This effect is unimportant in the ordinary methods of measuring α-particle ranges where the range is judged from the ionization in the last few millimetres of the track, but would have to be taken into account if some type of discharge counter sensitive to a few ions were used to determine the limits of tracks. The effect should be small, and while it has been commented on by one or two authors no quantitative information appears to have been published.

No conventional scale has been set up for particles other than α-particles, measurements of proton ranges are made assuming some ionization-range relation. An error as large as 1 mm. in a proton range amounts to only 5500 e.V. for an 80 cm. range, 9000 e.V. for a 30 cm. range, 15,000 e.V. for a 10 cm. range, 30,000 e.V. for a 2 cm. range. The form of the ionization-range relation (Fig. 1·1) is fairly certain up to within 1 mm. from the end of the range(22a), errors should therefore be only of minor importance. Geometrical considerations and the composition of the stopping medium are of much greater importance in determining the longer proton ranges. There is still some uncertainty about the range-energy relation for protons, as measurements of path curvature in a magnetic field combined with range measurements are wanting; the actual range-energy relation at present adopted is a theoretical extrapolation, the constants in the formula being derived from the experimental α-particle range-energy relation. The extrapolation is probably accurate, but the validity of the assumed constants appears less certain.

APPENDIX

It may be useful to give an example of the corrections applied to experimental measurements of an α-particle range to deduce the range on the conventional scale and hence the energy of the particles. This will serve to illustrate the order of magnitude of the several corrections.

Distance of ionization chamber from source corresponding to extrapolated numbers range ...   4·00 cm.

Penetration of particles into the chamber for this setting (determined in a separate experiment with a standard source)   ...   .   0·30 cm.

Stopping power of mica screen (calibrated for emergent α-particle range of 1·0 cm.)   ...   2·00 cm.

Correction for actual thickness of mica displacing air   ...   .   −0·001 cm.

Correction to stopping power of mica screen for emergent range of 6 mm. Not known accurately but probably not less than   ...   ...   −0·01 cm.

Room temperature 20° C. Barometer reading 750 mm.

Correction of −3·02 % to air distance, viz. 4·0 cm.   −0·121 cm.

(For the highest accuracy it is necessary to consider the effect of atmospheric density on the chamber characteristic, but usually it is sufficiently accurate to neglect the penetration distance.)

Correction of mercury column to 0° C. (2·44 mm. Hg)   ...   .   −0·014 cm.

Correction to sea level at latitude 45° (0·5 mm.)...   +0·003 cm.

Correction for 70 % humidity at 20° C. ...   ...   −0·017 cm.

Secant effect correction on 6·30 cm.   ...   ...   +0·025 cm.

Corrected extrapolated numbers range   ...   6·165 cm.

Straggling parameter α^2 for air   6·3 cm.   0·922

α^2 for mica 2·0 cm.   0·145

$-\alpha^2$ for air   2·0 cm.   0·117

$\Sigma\alpha^2$ 0·950

$\alpha = 0·974\,\text{mm.}$

Correction for mean range $\dfrac{-\alpha\sqrt{\pi}}{2} = 0·886\alpha$ ... $-0·086$ cm.

Mean range on conventional scale ...   ...   6·079 cm.

Having arrived at this final corrected mean range the corresponding energy is obtained from the range-energy curve [32,42] if this can be read with sufficient accuracy. If not reference must be made to the range velocity data [40]. A convenient difference curve may be constructed from these data [62,64].

To obtain the energy released in the disintegration allowance must be made for the energy of recoil of the residual nucleus, and if it is an artificial transmutation an appropriate correction for the energy of the bombarding particle.

REFERENCES

(1) H. Abraham and E. Bloch. *Annales de Phys.* **12** (1919), 237.
(2) H. Alfvén. *Nature, Lond.*, **136** (1935), 70, 394.
(3) H. Alfvén. *Proc. Phys. Soc.* **50** (1938), 358.
(4) H. Bateman. *Phil. Mag.* **20** (1910), 704.
(5) Z. Bay. *Nature, Lond.*, **141** (1938), 284, 1011.
(6) W. E. Bennett. *Proc. Roy. Soc.* A, **155** (1936), 419.
(7) H. Bethe. *Ann. Phys., Lpz.*, **5** (1930), 325.
(8) H. S. Black. *Elect. Engng*, **53** (1934), 114; reprinted in *Bell Syst. Tech. J.* **13** (1934), 1.
(9) F. Bloch. *Ann. Phys., Lpz.*, **16** (1933), 285; *Z. Phys.* **81** (1933), 363.
(10) A. D. Blumlein, C. O. Browne, N. E. Davis and E. Green. *J. Instn Elect. Engrs*, **83** (1938), 758.
(11) G. W. Bowdler. *J. Instn Elect. Engrs*, **73** (1933), 65.
(12) G. H. Briggs. *Proc. Roy. Soc.* A, **114** (1927), 313.
(13) G. H. Briggs. *Proc. Roy. Soc.* A, **139** (1933), 638; **143** (1934), 604.
(14) G. H. Briggs. *Proc. Roy. Soc.* A, **157** (1936), 183.
(15) Cenco High Speed Counter.
(16) Cosyns and de Bruyn. *Bull. Acad. Belg. Cl. Sci.* **20** (1934), 371. Cosyns. *Bull. tech. Ass. Ing. Brux.* (1936), p. 173.
(17) P. Debye. *Ann. Phys., Lpz.*, **39** (1912), 789.
(18) O. S. Duffendack, H. Lifschutz and M. M. Slawsky. *Phys. Rev.* **52** (1937), 1231.
(19) W. E. Duncanson. *Proc. Camb. Phil. Soc.* **30** (1934), 102.
(20) W. H. Eccles and F. W. Jordan. *Radio Rev.* **1** (1920), no. 3.
(21) R. D. Evans. *Rev. Sci. Instrum.* **5** (1934), 371.
(22) N. Feather and R. R. Nimmo. *Proc. Camb. Phil. Soc.* **24** (1928), 139.
(22a) N. Feather. *Nature, Lond.*, **147** (1941), 510.
(23) Flammersfeld. *Naturwissenschaften*, **24** (1936), 522.
(24) C. van Geel and J. Kerkum. *Physica*, **5** (1938), 609.
(25) H. Geiger and O. Klemperer. *Z. Phys.* **49** (1928), 753.
(26) I. A. Getting. *Phys. Rev.* **53** (1938), 103.
(27) N. S. Gingrich, R. D. Evans and H. E. Edgerton. *Rev. Sci. Instrum.* **7** (1936), 450.
(28) R. W. Gurney. *Proc. Roy. Soc.* A, **107** (1925), 340.
(29) C. A. Hartmann and H. Dossmann. *Z. Tech. Phys.* **9** (1928), 434.
(30) G. H. Henderson. *Proc. Roy. Soc.* A, **102** (1922), 496; **109** (1925), 157.

(31) E. W. Herold. *Proc. Inst. Radio Engrs, N.Y.,* **23** (1935), 1201.
(32) M. G. Holloway and M. S. Livingston. *Phys. Rev.* **54** (1938), 18.
(33) A. W. Hull. *Proc. Inst. Radio Engrs, N.Y.,* **6** (1918), 5.
(34) A. W. Hull. *Gen. Elect. Rev.* **32** (1929), 213, 390.
(35) W. Jentschke. *S.B. Akad. Wiss. Wien* (11*a*), **144** (1935), 151.
(36) J. B. Johnson. *Phys. Rev.* **26** (1925), 71.
(37) W. B. Lewis. *Proc. Camb. Phil. Soc.* **30** (1934), 543.
(38) W. B. Lewis. *Proc. Camb. Phil. Soc.* **33** (1937), 549.
(39) W. B. Lewis and B. V. Bowden. *Proc. Roy. Soc.* A, **145** (1934), 235.
(40) W. B. Lewis and C. E. Wynn-Williams. *Proc. Roy. Soc.* A, **136** (1932), 349.
(41) H. Lifschutz and J. L. Lawson. *Rev. Sci. Instrum.* **9** (1938), 83.
(42) M. S. Livingston and H. A. Bethe. *Rev. Mod. Phys.* **9** (1937), 245.
(43) F. Löhle. *Z. Phys. Chem. Unterr.* **46** (1933), 169.
(44) G. Mano. *J. Phys. Radium,* **5** (1934), 628.
(45) E. Marsden and H. Richardson. *Phil. Mag.* **25** (1913), 184.
(46) K. A. Macfadyen. *Wireless Engr,* **12** (1935), 639.
(47) A. N. May. *Proc. Phys. Soc.* **51** (1939), 26. Also *Reports on Progress in Physics,* **5** (1939), 390.
(48) R. Maze. *J. Phys. Radium,* **9** (1938), 162.
(49) E. B. Moullin. *Spontaneous Fluctuations of Voltage* (1938). Oxford: Clarendon Press.
(50) H. V. Neher. *Rev. Sci. Instrum.* **10** (1939), 29.
(51) H. V. Neher and W. W. Harper. *Phys. Rev.* **49** (1936), 940.
(52) Neon stabilizers are made by Cossor, Mullard and Marconi's Wireless Telegraph Co.
(53) H. Nyquist. *Phys. Rev.* **32** (1928), 110.
(54) G. L. Pearson, *Physics,* **6** (1935), 6.
(55) W. H. Rann. *Nature, Lond.,* **141** (1938), 410.
(56) H. J. Reich. *Rev. Sci. Instrum.* **9** (1938), 222.
(57) B. Rossi. *Nature, Lond.,* **125** (1930), 636.
(58) S. Rosenblum and G. Dupouy. *J. Phys. Radium,* **4** (1933), 262.
S. Rosenblum and C. Chamié. *C.R. Acad. Sci., Paris,* **196** (1933), 1663.
Mme P. Curie and S. Rosenblum. *C.R. Acad. Sci., Paris,* **196** (1933), 1598.
S. Rosenblum and M. Valadares. *C.R. Acad. Sci., Paris,* **194** (1932), 967.
S. Rosenblum, M. Guillot and M. Perey. *C.R. Acad. Sci., Paris,* **202** (1936), 1274.
S. Rosenblum. *C.R. Acad. Sci., Paris,* **195** (1932), 317; **193** (1931), 848.
(59) A. Ruark. *Phys. Rev.* **53** (1938), 316.

(60) Rutherford. *Phil. Mag.* **47** (1924), 277.

(61) Rutherford, Chadwick and Ellis. *Radiations from Radioactive Substances* (1930). Camb. Univ. Press.

(62) Rutherford, Ward and Lewis. *Proc. Roy. Soc.* A, **131** (1931), 684.

(63) Rutherford, Ward and Wynn-Williams. *Proc. Roy. Soc.* A, **129** (1930), 211.

(64) Rutherford, Wynn-Williams, Lewis and Bowden. *Proc. Roy. Soc.* A, **139** (1933), 617.

(65) H. Schulze. *Z. Phys.* **94** (1935), 104.

(66) W. G. Shepherd and R. O. Haxby. *Rev. Sci. Instrum.* **7** (1936), 425.

(67) E. S. Shire. *J. Sci. Instrum.* **11** (1934), 379.

(68) E.g. Standard Telephones Types 4007 and 4055.

(69) A. T. Starr. *Wireless Engr,* **12** (1935), 601. Gives a bibliography of this type of circuit.

(70) G. Stetter and W. Jentschke. *Phys. Z.* **36** (1935), 441.

(71) J. C. Street and T. H. Johnson. *J. Franklin Inst.* **214** (1932), 155.

(72) E. C. Stevenson and I. A. Getting. *Rev. Sci. Instrum.* **8** (1937), 414.

(73) D. J. Struik. *J. Math. Phys.* **9** (1929–30), 151. Gives an account of many derivations of Poisson's Law.

(74) A. Trost. *Z. Phys.* **105** (1937), 399.

(75) J. L. Tuck. *J. Sci. Intrum.* **13** (1936), 366.

(76) L. Ürmenyi. *Helv. Phys. Acta,* **10** (1937), 285.

(77) S. Werner. *Z. Phys.* **92** (1934), 705; **90** (1934), 384.

(78) L. Wertenstein. *Phys. Rev.* **54** (1938), 306.

(79) C. E. Wynn-Williams. Brit. Pat. No. 421341.

(80) C. E. Wynn-Williams. *Proc. Roy. Soc.* A, **136** (1932), 312.

(81) C. E. Wynn-Williams and F. A. B. Ward. *Proc. Roy. Soc.* A, **131** (1931), 391.

(82) B. Zipprich. *Z. Phys.* **85** (1933), 592.

Printed in the United States
By Bookmasters